PROCEEDINGS
SPIE—The International Society for Optical Engineering

Eyesafe Lasers: Components, Systems, and Applications

Anthony M. Johnson
Chair/Editor

21 January 1991
Los Angeles, California

Sponsored and Published by
SPIE—The International Society for Optical Engineering

SPIE PROCEEDINGS SERIES

Volume 1419

SPIE (Society of Photo-Optical Instrumentation Engineers) is a nonprofit society dedicated to the advancement of optical and optoelectronic applied science and technology.

Please use the following format to cite material from this book:
Author(s), "Title of Paper," *Eyesafe Lasers: Components, Systems, and Applications,* Anthony M. Johnson, Editor, Proc. SPIE 1419, page numbers (1991).

Library of Congress Catalog Card No. 91-60495
ISBN 0-8194-0509-4

Published by
SPIE—The International Society for Optical Engineering
P.O. Box 10, Bellingham, Washington 98227-0010 USA
Telephone 206/676-3290 (Pacific Time) • Fax 206/647-1445

Printed in the United States of America

EYESAFE LASERS: COMPONENTS, SYSTEMS, AND APPLICATIONS

Volume 1419

CONTENTS

(continued)

EYESAFE LASERS: COMPONENTS, SYSTEMS, AND APPLICATIONS

Volume 1419

EYESAFE LASERS: COMPONENTS, SYSTEMS, AND APPLICATIONS

Volume 1419

CONFERENCE COMMITTEE

Conference Chair
Anthony M. Johnson, Optic-Electronic Corporation

Cochairs
Penelope K. Galoff, U.S. Army Environmental Hygiene Agency
Milton A. Woodall II, Optic-Electronic Corporation

Session Chairs
Session 1—Eyesafe Laser Components
Penelope K. Galoff, U.S. Army Environmental Hygiene Agency
Milton A. Woodall II, Optic-Electronic Corporation

Session 2—Eyesafe Laser Systems and Applications
Milton A. Woodall II, Optic-Electronic Corporation
Penelope K. Galoff, U.S. Army Environmental Hygiene Agency

Conference 1419, *Eyesafe Lasers: Components, Systems, and Applications*, was part of a five-conference program on Laser Beam Applications and Diagnostics held at SPIE's Symposium on High-Power Lasers, a part of OE/LASE '91, 20–25 January 1991, in Los Angeles, California. The other conferences were:

Conference 1414, *Laser Beam Diagnostics*
Conference 1415, *Modeling and Simulation of Laser Systems II*
Conference 1416, *Laser Radar VI*
Conference 1417, *Free-Space Laser Communication Technologies III.*

Program Chair: **Richard J. Becherer,** Science Applications International Corporation

INTRODUCTION

Previous conferences addressed various aspects of lasers in reference to their potential ocular hazards; Session 1 of this year's conference presented a brief overview of the mechanisms involved with eye safety. A separate paper described how protective devices are used and tested to protect users from the possibility of eye damage. Other papers described key components used in eyesafe laser systems, with special emphasis on various indium gallium arsenide detectors and receiver subsystems.

Session 2 directed attention to eyesafe lasers and their application for fire control, military, and law enforcement training systems. A wide variety of types of laser systems were described, including gallium arsenide, Raman-shifted Nd:YAG, holmium, and carbon dioxide.

World-wide participation in the sessions demonstrated the growth of interest in laser eye safety. The quality of the papers presented and the outstanding attendance is a tribute to the individual authors and the efforts of cochairs Penelope K. Galoff and Milton A. Woodall II. Collection and submission of the manuscripts is due to the efforts of Ms. Patti Weekly, who acted as the session administrative coordinator.

Anthony M. Johnson
Optic-Electronic Corporation
Varo/OEC Systems Division of Imo Industries, Inc.

SESSION 1

Eyesafe Laser Components

Chairs
Penelope K. Galoff
U.S. Army Environmental Hygiene Agency

Milton A. Woodall II
Optic-Electronic Corporation

What is eyesafe?

James K. Franks*

U. S. Army Environmental Hygiene Agency
Aberdeen Proving Ground, Maryland 21010-5422

ABSTRACT

Manufacturers have referred to lasers with operating wavelengths longer than 1400 nm (mid to far infrared) as eyesafe. Wavelengths in this region are absorbed in anterior portions of the eye (mainly cornea) and therefore never reach the retina. This is in contrast to the eye-hazardous portion of the optical spectrum of 400 - 1400 nm (visible and near infrared) where the anterior portions of the eye have high transmittance and refractive power. Irradiance levels are typically 5 orders of magnitude greater at the retina than at the cornea for visible and NIR wavelengths. Although wavelengths longer than 1400 nm do not interact with the retina, they can interact with the skin or cornea and cause a thermal injury.

Specifiers of laser equipment have required that systems be eyesafe at a particular distance with or without optics. By this they mean that exposure at that distance for the naked eye or optically aided viewing through a specified power of optics should not exceed exposure limits. Some wavelengths in the visible and near IR are more "eyesafe" than others because the safety factor between the level that would actually produce an injury (ED-50 value) and the exposure limit is greater. For example, the ED-50 value for a single Q-switched laser pulse for the ruby laser wavelength of 694.3 nm is 11.2 μJ into the eye[1] and the Accessible Emission Limit (AEL) is 0.19 μJ, a factor of 59 between ED-50 and the AEL. The ratio for 1064 nm for a single Q-switched pulse is 99 μJ/1.9 μJ or 52. So, in a sense, exposure to 694 laser radiation at the AEL, is slightly more "eyesafe" than exposure to 1064 nm radiation at the AEL.

To the laser safety specialist, however, eyesafe, means only one thing, Class 1. Class 1 is a laser hazard classification that means the laser does not exceed specified limits for a reasonable, worst-case combination of distance, exposure duration, and collecting aperture diameter during its intended use. Determining whether a laser is Class 1 is usually a two-step process. The first step is determination of the AEL and the second is determining whether or not the laser in question exceeds this AEL. Exceeding the AEL is determined by calculation or measurement. If the AEL is exceeded under prescribed parameters then the laser is not Class 1. Differences in laser safety standards in both of these classification steps have caused confusion. Differences in exposure limits chosen by laser safety specialists based upon the same criteria have also caused confusion.

*The opinions or assertions contained herein are the private views of the author and do not necessarily reflect the official views of the Department of the Army or the Department of Defense.

1. CLASSIFICATION STANDARDS

1.1. TB MED 524

The document that the Army has used to classify lasers until recently is TB MED 524, Control of Hazards to Health from Laser Radiation.[2] This document was last updated in 1985 and defines Class 1 as:

"Any laser device that cannot emit laser radiation levels in excess of the AEL for the maximum possible duration inherent in the design of the laser or laser system. The AEL is the accessible emission limit and is usually the product of the intrabeam exposure limit for the limiting exposure duration and the circular area of the defining aperture."

The circular area of the defining aperture is the easy part. The circular area for the defining apertures are given in Table 1.

TABLE 1. Defining Aperture Areas for TB MED 524-1985 and ANSI Z136.1-1986.

Wavelength Range	Defining Aperture(cm)	Area(cm^2)
>400 nm to 1400 nm	0.7	0.385
>1400 nm to 0.1 mm and >200 nm to 400 nm	0.1	0.00785
>0.1 mm to 1mm	1.1	1.0

Exposure limits for typical "eyesafe" lasers are given in Table 2.

TABLE 2. Exposure Limits (EL) and Accessible Emission Limits (AEL) for Typical "Eyesafe Lasers" from TB MED 524-1985 and ANSI Z136.1-1986 AEL(1) and FDA AEL(2).

Laser-Use	Wavelength	PRF	Assumed Exposure Duration	EL	AEL(1)	AEL(2)
CO_2-LRF	10,600	1	20 ns	10 mJ/cm^2	78.5 μJ	10 mJ/cm^2
CO_2 Beamrider	10,600	cw	10 s or longer	100 mW/cm^2	785 μW	100 mW/cm^2
Shifted YAG LRF	1,540	4-10	10 s	4 Hz - 40 mW/cm^2 10 Hz - 32 mW/cm^2	314 μW 251 μW	780 μW 780 μW
Erbium Glass LRF	1,540	1	25 ns	1 J/cm^2	7.9 mJ	7.9 mJ

As you can see, exposure limits and therefore hazard classification depend upon the exposure duration chosen. Typically 8 hours is chosen for cw lasers if no system safety features limit this. For single-pulse lasers, exposure duration is assumed to be the laser pulse width. For repetativelypulsed IR lasers, the accidental exposure duration is 10 s. In the following example, is the laser Class 1 according to the Army standard?

Laser Type---Carbon dioxide laser rangefinder
Wavelength---10,600 nm
Pulse repetition frequency----single shot
Pulse width------20 ns
Radiant Energy----80 mJ
Beam diameter----5 cm at 1/e points in the Gaussian profile.

STEP 1
Determine the AEL. From Table 2 above the exposure limit for a single-pulse carbon dioxide laser with a 20 ns pulse width is 10 mJ/cm^2. The area of the defining aperture is 7.85×10^{-3} cm^2. The AEL is the product of these numbers or 78.5 μJ.

STEP 2
Is the AEL exceeded? The answer is clearly yes since the total output is 80 mJ this radiant energy clearly exceeds the AEL of 78.5 μJ. The TB MED does not specify an aperture over which to collect power, it implies that you must collect the total radiant output. Hence this laser is not Class 1 according to TB MED 524.

1.2. Food and Drug Administration (FDA) Regulation-21 CFR 1040

The Food and Drug Administration requires that all lasers (except certain kinds of military lasers) manufactured or sold in the United States since August of 1986 have certain system safety features built into them depending on their laser hazard classification.[3] This regulation defines a Class 1 laser product as:

"Any laser product that does not permit access during the operation to levels of laser radiation in excess of the accessible emission limits given in Table I of paragraph (d) of this section."

Table I from the FDA rule is given as Figure 1. The AEL's according to the FDA rule for typical "eyesafe" lasers are given in Table 2. The Class 1 AEL for our example is 10 mJ/cm^2. Notice that these AEL's are not given as the product of an exposure limit and the circular area of the defining aperture as for TB MED 524. Also note the following: The only laser that has the same AEL as the TB MED is the 1.54 μm single-pulse LRF. Pulse-repetition-frequency does not effect AEL as it does for the TB MED and the ANSI standard.

We have finished step 1 in our 2-step process. To finish step 2 let's go back to our example. To see if the laser was Class 1 using the TB MED we compared the total output of the laser with the AEL. The FDA rule requires that the output parameters of the laser be measured or calculated under specific conditions described in 21 CFR 1040[2]. In the FDA regulation, AEL that are given in units of power or energy must be is collected through a 7mm aperture unless optics are used, then, a 50mm collecting aperture is required. FDA AEL that are given in units of irradiance or radiant exposure require that the measurement be averaged over a 7mm detecting aperture. In our example the FDA-AEL is given as 10 mJ/cm^2. If we measured the radiant exposure at the laser exit optics, assuming that the beam is Gaussian and we average over a 7 mm aperture, we should get a value very similar to The radiant energy divided by the area of the effective beam given above or, 80 $mJ/19.6\ cm^2$. This equals $4.1\ mJ/cm^2$, a level a factor of 2 below the AEL.

Our laser is Class 1 according to the FDA.

1.3. ANSI STANDARD

The American National Standard for the Safe Use of Lasers[4] was last revised in 1986 and is undergoing a current revision. This standard is a consensus standard and as the title suggests, it is for the user of laser systems. The standard relies heavily on the judgement of someone called a Laser Safety Officer (LSO) and gives this knowlegeable individual latitude in laser hazard classification. We already know that except for certain kinds of military lasers, lasers sold in commerce in the United States will already have been classified by the manufacturer. The LSO however, because of the individual laser use, may classify the laser differently than FDA. He could give a laser a lower classification based upon the fact that the exposure is limited to a shorter duration than used by the FDA in their product performance standard. The ANSI-1986 defines Class 1 the same as the TB MED above except that the LSO is given more latitude in the ANSI standard as to maximum exposure duration. AEL is defined the same as in the TB MED, i. e. the AEL is the product of the exposure limit and the area of the defining aperture. The defining apertures are defined in the same way. The difference is in what laser parameter you use to compare to the AEL. the TB MED requires that you measure the total output of the laser regardless of beam size, while the ANSI standard defines the collecting apertures as in Figure 2. The standard says to use 50 mm collecting optics if it is likely that optics will be used for intrabeam viewing, if it is unlikely, then the limiting aperture size is to be used. In our example, the wavelength is such that glass optics will not transmit it. The standard then says to use the listed limiting apertures given in the table to collect the laser output. Assuming a Gaussian distribution in the beam profile, the amount of power that would pass through an aperture 1mm in diameter is (from Equation 10, TB MED 524-1985) $32\ \mu J$. The AEL is the same defined in Table 2 or 78.5 μJ. Therefore since the amount of power that would pass through a 1 mm aperture is less than the AEL, use of the ANSI standard results hazard Class 1 for our example.

1.3. MIL STD 1425.

Another standard, MIL STD 1425, Safety Design Requirements for Military Lasers and Associated Support Equipment, 13 December 1983, could also be used to classify a laser. This document applies to military lasers and implements an agreement between the DOD and the FDA for exemption of certain kinds of military lasers from the FDA rule described above. The military standard requires that the ANSI standard be used for classification of lasers that have been granted a certificate of exemption from the FDA rule by the DOD. The standard requires use of the FDA classification scheme for those lasers that are not granted this certificate of exemption. Therefore for our example, if the system were exempted (it would normally have been), ANSI would arrive at Class 1. If it had not been exempted it would still be Class 1 according to the FDA.

2. CONCLUSIONS

1. It is to the advantage of the DOD to not issue certificates of exemption for lasers that can meet the FDA Class 1 criteria.

2. The FDA rule has some advantages over the other standards. For example, it uses only a radiant exposure criteria for wavelengths longer than 2500 nm. This seems to be incorporating the transmission of glass directly into the standard. An improvement would be to use 3000 nm rather than 2500 nm as the cut-off for glass transmission.

3. The FDA standard has some weaknesses. It does not use a PRF correction factor to determine exposure limit. This factor has been shown to have a basis in biological effect even for far infrared wavelengths. A measurement aperture of 1 mm would be better than 7 mm for far infrared wavelengths since hotspots of this diameter are biologically significant.

3. REFERENCES

1. David Lund and Edwin S. Beatrice, "Near Infrared Laser Ocular Effects,"Health Physics, Vol 56, No 5, (May), pp 631-636, 1989.

2. Department of the Army, Control of hazards to health from laser radiation. Washinton, DC: Department of the Army; TB MED 524, 20 June 1985.

3. U.S. Food and Drug Administration, Performance standard for laser products. Washington, DC: U.S. Government Printing Office; 21 Code of Federal Regulations 1040.10(e)(3); 1987.

4. American National Standards Institute, American national standard for the safe use of lasers, New York, NY: ANSI, 1430 Broadway; ANSI Z136.1-1986; 1986.

Wavelength (nanometers)	Emission duration (seconds)	Class I-Accessible emission limits (value)	(unit)	(quantity)**
≥180 but ≤400	≤3.0 X 10^4	2.4 X $10^{-5}k_1k_2$*	Joules(J)*	radiant energy
	>3.0 X 10^4	8.0 X $10^{-10}k_1k_2$*	Watts(W)*	radiant power
>400 but ≤1400	>1.0 X 10^{-9} to 2.0 X 10^{-5}	2.0 X $10^{-7}k_1k_2$	J	radiant energy
	>2.0 X 10^{-5} to 1.0 X 10^1	7.0 X $10^{-4}k_1k_2t^{3/4}$	J	radiant energy
	>1.0 X 10^1 to 1.0 X 10^4	3.9 X $10^{-3}k_1k_2$	J	radiant energy
	>1.0 X 10^4	3.9 X $10^{-7}k_1k_2$	W	radiant power
	and also (See paragraph (d)(4) of this section)			
	>1.0 X 10^{-9} to 1.0 X 10^1	$10k_1k_2t^{1/3}$	$Jcm^{-2}sr^{-1}$	integrated radiance
	>1.0 X 10^1 to 1.0 X 10^4	$20k_1k_2$	$Jcm^{-2}sr^{-1}$	integrated radiance
	>1.0 X 10^4	2.0 X $10^{-3}k_1k_2$	$Wcm^{-2}sr^{-1}$	radiance
>1400 but ≤2500	>1.0 X 10^{-9} to 1.0 X 10^{-7}	7.9 X $10^{-5}k_1k_2$	J	radiant energy
	>1.0 X 10^{-7} to 1.0 X 10^1	4.4 X $10^{-3}k_1k_2t^{1/4}$	J	radiant energy
	>1.0 X 10^1	7.9 X $10^{-4}k_1k_2$	W	radiant power
>2500 but ≤1.0 X 10^6	>1.0 X 10^{-9} to 1.0 X 10^{-7}	1.0 X $10^{-2}k_1k_2$	Jcm^{-2}	radiant exposure
	>1.0 X 10^{-7} to 1.0 X 10^1	5.6 X $10^{-1}k_1k_2t^{1/4}$	Jcm^{-2}	radiant exposure
	>1.0 X 10^1	1.0 X $10^{-1}k_1k_2t$	Jcm^{-2}	radiant exposure

*Class I accessible emission limits for wavelengths equal to or greater than 180 nm but less than or equal to 400 nm shall not exceed the Class I accessible emission limits for the wavelengths greater than 1400 nm but less than or equal to 1.0 X 10^6 nm with a k_1 and k_2 of 1.0 for comparable sampling intervals.

**Measurement parameters and test conditions shall be in accordance with paragraphs (d)(1), (2), (3), and (4), and (e) of this section.

Figure 1. Accessible Emission Limits (EL) for FDA Class 1 Laser Products.

Maximum Aperture Diameters (Limiting Aperture) for Measurement Averaging

		Wavelength Range			
Measurement	Exposure Duration, t (s)	Ultraviolet (0.2 to 0.4 μm)	Visible and Near Infrared (0.4 to 1.4 μm)	Medium and Far Infrared (1.4 to 10^2 μm)	Submillimeter (0.1 to 1 mm)
Eye MPE	10^{-9} to 3×10^4	1 mm	7 mm	1 mm	11 mm
Skin MPE	10^{-9} to 3×10^4	1 mm	1 mm	1 mm	11 mm
Laser Classification*	10^{-9} to 3×10^4	50 mm	50 mm	50 mm	50 mm

* The apertures are used for the measurement of total output power or output energy for laser classification purposes, that is, to distinguish between all classes of cw lasers or between Class 1 and Class 3 pulsed lasers. The use of the 50-mm apertures as shown in the horizontal line labeled "Laser Classification" applies only to those cases where the laser output is intended to be viewed with optical instruments (excluding ordinary eyeglass lenses) or where the Laser Safety Officer determines that there is some probability that the output will be accidentally viewed with optical instruments and that such radiation will be viewed for a sufficient time duration so as to constitute a hazard. Otherwise the apertures listed for Eye MPE and Skin MPE are to be used.

For the specific case of optical viewing (beam collecting) instruments, the apertures listed for eye MPE and skin MPE apply to the exit beam of such devices.

Figure 2. ANSI Classification Aperture Diameters.

Wavelength (nanometers)	Emission duration (seconds)	Class I-Accessible emission limits		
		(value)	(unit)	(quantity)**
≥ 180 but ≤ 400	$\leq 3.0 \times 10^{4}$	$2.4 \times 10^{-5} k_1 k_2$*	Joules(J)*	radiant energy
	$>3.0 \times 10^{4}$	$8.0 \times 10^{-10} k_1 k_2$*	Watts(W)*	radiant power
>400 but ≤ 1400	$>1.0 \times 10^{-9}$ to 2.0×10^{-5}	$2.0 \times 10^{-7} k_1 k_2$	J	radiant energy
	$>2.0 \times 10^{-5}$ to 1.0×10^{1}	$7.0 \times 10^{-4} k_1 k_2 t^{3/4}$	J	radiant energy
	$>1.0 \times 10^{1}$ to 1.0×10^{4}	$3.9 \times 10^{-3} k_1 k_2$	J	radiant energy
	$>1.0 \times 10^{4}$	$3.9 \times 10^{-7} k_1 k_2$	W	radiant power
	and also (See paragraph (d)(4) of this section)			
	$>1.0 \times 10^{-9}$ to 1.0×10^{1}	$10 k_1 k_2 t^{1/3}$	$Jcm^{-2}sr^{-1}$	integrated radiance
	$>1.0 \times 10^{1}$ to 1.0×10^{4}	$20 k_1 k_2$	$Jcm^{-2}sr^{-1}$	integrated radiance
	$>1.0 \times 10^{4}$	$2.0 \times 10^{-3} k_1 k_2$	$Wcm^{-2}sr^{-1}$	radiance
>1400 but ≤ 2500	$>1.0 \times 10^{-9}$ to 1.0×10^{-7}	$7.9 \times 10^{-5} k_1 k_2$	J	radiant energy
	$>1.0 \times 10^{-7}$ to 1.0×10^{1}	$4.4 \times 10^{-3} k_1 k_2 t^{1/4}$	J	radiant energy
	$>1.0 \times 10^{1}$	$7.9 \times 10^{-4} k_1 k_2$	W	radiant power
>2500 but $\leq 1.0 \times 10^{6}$	$>1.0 \times 10^{-9}$ to 1.0×10^{-7}	$1.0 \times 10^{-2} k_1 k_2$	Jcm^{-2}	radiant exposure
	$>1.0 \times 10^{-7}$ to 1.0×10^{1}	$5.6 \times 10^{-1} k_1 k_2 t^{1/4}$	Jcm^{-2}	radiant exposure
	$>1.0 \times 10^{1}$	$1.0 \times 10^{-1} k_1 k_2 t$	Jcm^{-2}	radiant exposure

*Class I accessible emission limits for wavelengths equal to or greater than 180 nm but less than or equal to 400 nm shall not exceed the Class I accessible emission limits for the wavelengths greater than 1400 nm but less than or equal to 1.0×10^{6} nm with a k_1 and k_2 of 1.0 for comparable sampling intervals.

**Measurement parameters and test conditions shall be in accordance with paragraphs (d)(1), (2), (3), and (4), and (e) of this section.

Figure 1. Accessible Emission Limits (EL) for FDA Class 1 Laser Products.

Maximum Aperture Diameters (Limiting Aperture) for Measurement Averaging

Measurement	Exposure Duration, t (s)	Wavelength Range Ultraviolet (0.2 to 0.4 μm)	Visible and Near Infrared (0.4 to 1.4 μm)	Medium and Far Infrared (1.4 to 10^2 μm)	Submillimeter (0.1 to 1 mm)
Eye MPE	10^{-9} to 3×10^4	1 mm	7 mm	1 mm	11 mm
Skin MPE	10^{-9} to 3×10^4	1 mm	1 mm	1 mm	11 mm
Laser Classification*	10^{-9} to 3×10^4	50 mm	50 mm	50 mm	50 mm

* The apertures are used for the measurement of total output power or output energy for laser classification purposes, that is, to distinguish between all classes of cw lasers or between Class 1 and Class 3 pulsed lasers. The use of the 50-mm apertures as shown in the horizontal line labeled "Laser Classification" applies only to those cases where the laser output is intended to be viewed with optical instruments (excluding ordinary eyeglass lenses) or where the Laser Safety Officer determines that there is some probability that the output will be accidentally viewed with optical instruments and that such radiation will be viewed for a sufficient time duration so as to constitute a hazard. Otherwise the apertures listed for Eye MPE and Skin MPE are to be used.

For the specific case of optical viewing (beam collecting) instruments, the apertures listed for eye MPE and skin MPE apply to the exit beam of such devices.

Figure 2. ANSI Classification Aperture Diameters.

Receivers for eyesafe laser rangefinders--an overview

Ian D. Crawford

Analog Modules, Inc.
126 Baywood Avenue, Longwood, Florida 32750

ABSTRACT

High performance receivers allow lower-power, smaller lasers and assist in size reduction of a new generation of eyesafe laser rangefinders. Aspects of receiver design and trade-offs are covered to assist systems and application engineers.

1. BACKGROUND TECHNOLOGY

Time-of-flight laser rangefinders have previously used eye-hazardous wavelengths such as 0.69μm (ruby) or 1.06μm (YAG). The move toward 1.54μm for eye-safety reasons has resulted in changes in the detector and receiver design technology. 1.54μm lasers are heavier and less efficient than YAG, and simultaneously a desire for portability has resulted in a need for small size and low weight. Laser receiver design can help in two ways. By maintaining a high sensitivity, the laser output-energy and size may be minimized, and miniaturization of the receiver itself allows for direct savings and packaging flexibility. The most common detector for YAG has been the silicon avalanche photodetector (APD) type which provides up to 35A/W response. Because of the detector gain, preamplifier designs for YAG rangefinders have been optimized for overload recovery and damage protection rather than for the lowest noise. With the move to 1.54μm eyesafe lasers, germanium and InGaAs detectors are used to replace silicon which cuts off at 1.1μm. For operation over the wide temperature range usually required, InGaAs is the material of choice due to its lower leakage currents. The original price advantage of germanium has been greatly eroded as the yields of InGaAs detectors improve. *Table I* shows a comparison between germanium and InGaAs photodetectors. Up till now, PIN photodiodes have been preferred due to lower cost, the absence of high voltage bias circuits, less laser damage susceptibility, and larger area facilitating optical design. Germanium and InGaAs avalanche detectors are both available but none of the candidate detectors approaches the responsivity of the silicon APD, requiring the use of a low noise preamplifier. The additional amplifier gain makes layout, shielding, and EMI control more critical in eyesafe rangefinders. *Table 2* compares the features of PIN and APD photodetectors.

Solar background noise is less at 1.54μm and typically a lower wavelength blocking filter can replace an interference filter used in a 1.06μm receiver. Atmospheric backscatter is also lower and less time-programmed-gain reduction can be used at short ranges to allow ranging to difficult targets without false alarms or blocking due to backscatter. This does require a large instantaneous dynamic range capability to be designed into the receiver electronics.

TABLE 1
Germanium versus InGaAs photodetectors
Photodetectors

	Germanium	InGaAs
Cost	Low	Low-medium
Response	0.7A/W	0.8A/W
Size	Almost unlimited	Up to 3mm diameter
Capacitance,C	6pf (0.3mm diameter)	6pf (0.3mm diameter)
Speed	≈2nS, varies with manufacturer	2nS (0.3mm diameter)
Bias	No major difference	
Noise	0.5 to 1pA/√Hz (0.3mm diameter)	0.2pA/√Hz
Leakage Current	1μA, temperature and size dependent	30nA

TABLE 2
PIN detectors versus APD detectors
InGaAs Detectors

	PIN	APD
Cost	Low-medium	High
Response	0.8A/W	Up to 10A/W
Size	Up to 3mm	Up to 85μm
Capacitance,C	6pf (0.3mm diameter)	<1pf (85μm diameter)
Speed	2nS	<1nS
Bias	0 to 15V (affects C) Not critical	Up to 95V (affects response) Critical at high gains Temperature dependent
Damage	Good	Fair
Noise	0.2pA/√Hz	1.5pA/√Hz
Leakage Current	30nA (0.3μm diameter)	200nA (85μm diameter)

A new generation of laser receiver has been packaged into a three layer hybrid contained in a hermetic TO8 can. This design has been selected by the U.S. Army for a hand-held eyesafe rangefinder. Further development work is proceeding to evaluate the use of APD's. Preliminary results indicate a useful sensitivity increase from 40nW to <10nW. Evaluation of laser damage and other performance factors is currently underway.

2. DESIGN OF LASER RANGEFINDER RECEIVERS

Laser rangefinder receivers are designed specially for the purpose. They are different from fiber optic (FO) receivers, LIDAR receivers, and atmospheric communication receivers. Within the laser rangefinder receiver family, there are subsets for; higher range accuracy using a constant fraction discriminator (CFD); averaging multiple pulses; phase detection rangefinders; high power laser/designators; combined tracker/rangefinders; ground-to-air applications; and high PRF semiconductor lasers--to mention a few. To illustrate some of the differences, *Table 3* compares features of a typical FO receiver with a rangefinder receiver.

TABLE 3
Differences between
Laser rangefinder receivers and fiber optic receivers

Fiber Optic Receivers	Laser Rangefinder Receivers
■ Detector area optimized to fiber.	■ Detector area sets field of view (FOV) in conjunction with the optics
■ Detector overload and damage not normal	■ Detector damage threshold to be high ■ Amplifier must not damage when detector overloads ■ Both must recover quickly from the overload condition
■ 20-30dB optical dynamic range	■ Up to 60dB optical dynamic range
■ Low background sunlight	■ Must accommodate background sunlight
■ Multiple data pulses available to set gain control	■ Often a single data pulse with $1/R^2$ to $1/R^4$ time-gain law
■ Data word is usually of constant amplitude	■ Must respond to a second pulse of widely different amplitude. ■ Must not generate any additional spurious pulses after a strong pulse

A block diagram of a typical time-of-flight rangefinder receiver as shown in ***Figure 1*** described below might be used with an eyesafe laser.

2.1 Photodetector photodiode bias supply

Germanium and InGaAs PIN diodes require only a low voltage which is not critical. This is easily obtained from the dc power via an R-C filter.

APD's need a higher voltage, typically 70V, which requires a dc-dc converter or a feed from the pulse-forming-network. The latter approach is simpler especially if the APD bias controller is located within the laser receiver. It is desirable that switching converters are not operated during the receive period to reduce EMI. A number of methods of APD bias control have been used to maintain a constant gain and False Alarm Rate (FAR) despite temperature variations, namely;

A. Measure temperature and set the bias voltage to track the desired temperature coefficient;
B. Adjust the APD bias until the total noise increases by a measured amount;
C. Use a "blind" matching APD detector to set the bias;
D. Let the APD avalanche at a controlled current to determine the avalanche voltage, then pull back by a fixed voltage just prior to use to set the gain;
E. Use a fixed bias voltage and control the APD temperature;
F. Use a voltage-source to self-heat the APD to meet a preset avalanche point, then pull back by a known voltage before use.

Each of these methods has advantages and disadvantages. Our selection was made based on the degree of bias control accuracy needed.

The claimed temperature coefficient (T_C) of an InGaAs APD is +0.18V/°C for constant gain. Tests with a complete receiver with the APD bias adjusted for a constant FAR indicate a lower T_C is required, namely 0.08V/°C. The temperature coefficient of responsivity for the complete receiver was <0.6%/°C, not significant compared to the dynamic range of a laser rangefinder receiver. Bias method **A** was selected due to the relatively low value of T_C and the lack of constraints such as timing signals needed by some of the other methods. By incorporating the temperature sensor and HV regulator within the laser receiver, a simple external power source such as the laser transmitter pulse-forming-network (PFN) can be used.

2.2 Protection

Laser receivers can be damaged due to specular reflections caused, for example, during alignment testing. Two separate problems exist--detector damage and preamplifier damage.

Detector damage falls into two categories: a) Direct material damage and; b) Damage caused by light-induced electrical overheating.

FIG. 1 LASER RECEIVER BLOCK DIAGRAM

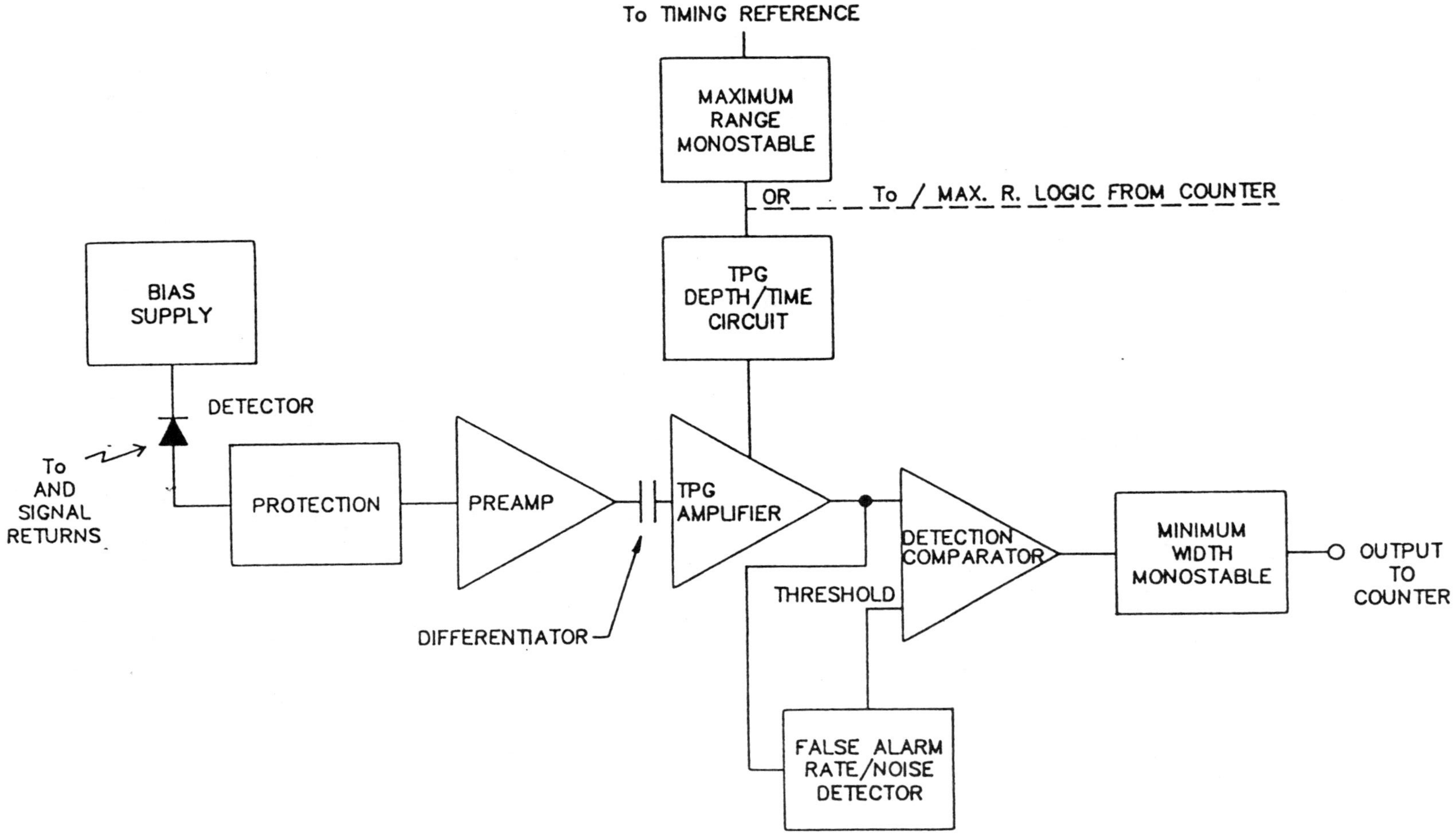

To TIMING REFERENCE
MAXIMUM RANGE MONOSTABLE
OR
To / MAX. R. LOGIC FROM COUNTER
TPG DEPTH/TIME CIRCUIT
BIAS SUPPLY
DETECTOR
To AND SIGNAL RETURNS
PROTECTION
PREAMP
TPG AMPLIFIER
DIFFERENTIATOR
DETECTION COMPARATOR
THRESHOLD
FALSE ALARM RATE/NOISE DETECTOR
MINIMUM WIDTH MONOSTABLE
OUTPUT TO COUNTER

To be material-limited is the goal. Even if the light is spread equally over the detector area, at some point the material will overheat and damage. Sunlight should be optically filtered so that it does not cause detector degradation directly. Laser pulses are more problematical since their short time can result in local instantaneous heating, especially if focussed to a point. Defocussing the spot over a large detector area can mitigate this problem, but increased capacitance results in reduced sensitivity. Electrical overheating damage is more likely in APD's. The light causes an electrical current to flow and the combination of relatively high voltage (≈70V), potentially high peak current, and a local hot spot can be damaging. The current may be limited by a series resistor, but this results in a de-biassing of the detector which may need to recover quickly, for example, to achieve a minimum range requirement. The resistor also degrades sensitivity by causing photocurrent to flow locally into the detector's self-capacitance. Recovery is also complicated by the recharging of the detector capacitance causing an exponential current to flow into the high-gain receiver. Another approach is to limit the energy stored within the detector decoupling capacitor by minimizing its size. The decoupling capacitor value must be much larger than the detector capacitance to avoid signal loss. Again the recharging of this capacitance causes exponential input currents to flow which are greater and faster with small values of capacitance.

All the above effects are worse with the APD because of its higher responsivity, high voltage bias, voltage-gain sensitivity, and smaller size.

Preamplifier damage is caused when the input ratings are exceeded and generally the input transistor is blown. This problem is invariably caused by the high voltage bias needed by an APD. When flooded with light, the APD exhibits a low series resistance and tries to directly apply the high voltage to the amplifier input. Protection via diodes and resistors is possible, but this can affect bandwidth and sensitivity because of the critical nature of the input of a high gain low-noise amplifier. On the other hand, Germanium or InGaAs PIN diodes require only a few volts of bias, well within the rating the input devices. Failure analysis of laser-damaged PIN receivers over a number of years confirm that amplifier damage is virtually unknown.

In summary, there is no simple solution. Protection trade-offs must be made against other parameters in a series of iterations to achieve the system goals.

2.3 Preamplifier

The preamplifier is designed to have low-noise, fast recovery from overload, and resistance to damage as described above. In addition, it must not saturate from the dc photocurrent caused by background illumination on the detector.

The low-noise requirement can only be met by using a high-gain preamplifier so that the thermal current noise from the feedback or load resistor is low. Unfortunately, this high impedance increases the recovery time and reduces the dynamic range of linear operation. Many manufacturers consider details of how they overcome this problem to be proprietary information, but in general, two type of preamplifier are used. The transimpedance virtual-ground input inverting type, and the voltage follower non-inverting buffer type. Both have advantages and disadvantages. The follower type is more manageable, easier to compensate, easier to bias, and has low noise. Capacitance of the detector and follower transistor is important and must be low to achieve good bandwidth. Layout is critical and generally this design is limited to small-area low-capacitance detectors. The transimpedance type can be more tolerant of detector capacitance if correctly compensated, but can give problems in an overload situation. The virtual ground must be preserved as well as possible to minimize recovery transients. One of the most difficult problems in designing a laser receiver preamplifier is to avoid any perturbations after recovery from a strong pulse which could cause a second pulse to cross threshold and give erroneous ranges if last-pulse-logic is used in the range counter. Similar design problems exist throughout the receiver but they are most critical in the preamplifier because the gain following can be up to 500 times.

2.4 Differentiation

The preamplifier is normally dc-coupled to the detector since ac-coupling would require another resistor connected to the input to sink the detector leakage and background light currents. This extra resistor would generate additional thermal noise and excess noise due to current flow. To prevent the dc light bias from saturating the post-amplifier stages, ac-coupling is normally used out of the preamplifier.

The value of this differentiation time-constant is selected to eliminate slow exponential backscatter components and to match the receiver low frequency response to the power spectrum of the laser pulse. It is important that this is the dominant LF cut-off since multiple time-constants lead to ringing and multiple false returns.

2.5. Time programmed gain (TPG) amplifier

Since a ranging operation is frequently a single event, automatic gain control can only be used in repetition rated systems--less common at eyesafe wavelengths. What is known is the requirement to increase sensitivity with time elapsed after the laser pulse is emitted (T_0). In a clear atmosphere, this law is $1/R^2$ when the target fills the beam, or $1/R^4$ when the target is smaller than the beam. Since the beam normally starts at 1-3" diameter, the $1/R^2$ law usually applies during the first part of the ranging time-of-flight. When the target is small and at long range, high gain is needed. For these reasons, the $1/R^2$ law is more desirable and the receiver is allowed to overload at short ranges.

3. ACKNOWLEDGMENTS

U.S. Army and Optic Electronic Corp., Dallas, Texas, supported parts of this work, and provided feedback from field tests.

4. REFERENCES

1. Epitaxx InGaAs photodetector data sheets.
2. GE (formerly RCA) Canada InGaAs photodetector data sheets.

Planar InGaAs APD for eye-safe laser rangefinding applications

Paul P. Webb

General Electric Canada Inc., Electro Optics Operations
P.O. Box 900, Vaudreuil, Quebec, Canada J7V 8P7

ABSTRACT

Planar InGaAs/InP avalanche photodiodes (APD's) for use in eye-safe laser rangefinding applications in the 1100 to 1700 nm spectral range are discussed. The devices have diameters of 85 and 200 μm, having capacitances of 0.7 and 2 pF, respectively. For the diodes operating at a responsivity of 10 A/W at 1540 nm, for which the gain is approximately 10, measured noise currents at room temperature are 0.5 and 0.75 pA/$\sqrt{}$Hz, respectively, and frequency response is greater than 400 MHz.

Results are also reported for a back-entry version of the 200 μm diameter APD fitted with a 600 μm diameter high index ball lens for enlarging the effective sensitive diameter. Quantum efficiency for the unit is found to be about 65%, and good response uniformity is achieved over a diameter greater than 500 μm, using an f/1.8 optical system.

1. INTRODUCTION

In recent years, the eye-safety hazard presented by the use of laser rangefinders operating in the wavelength range between about 900 and 1060 nm, has resulted in considerable interest in systems which operate at longer wavelengths, for which the eye-safety hazard is significantly reduced. The wavelength chosen for most current developments has been 1540 nm. PIN and APD detectors fabricated using the lattice-matched composition of InGaAs on InP, having a cutoff wavelength of about 1700 nm, and now being routinely fabricated for the fiber-optics communications industry, have good quantum efficiency at the required wavelength, and therefore are suitable candidates for the rangefinding application. It is the purpose of this paper to describe the design and performance characteristics of InGaAs/InP APD's which are considered to be useful for the rangefinding application. Detector sizes are 85 and 200 μm diameter, and the use of a ball lens mounted on a back-entry version of the 200 μm diameter device is shown to expand its effective diameter up to at least 500 μm for some applications.

2. BACKGROUND

The design and performance characteristics of an InGaAs/InP APD fabricated using ion implantation and regrowth techniques have been described previously.[1,2] In this design, the epitaxial layers are grown nominally undoped, and the carriers required to create the high-field multiplying region are introduced by implantation of silicon in a defined region. The implantation is followed by a second epitaxial growth step to deposit the capping layer in which the junction is formed. The device structure, impurity profile, and field configuration are shown in Figures 1 and 2. The basic design is the Separated Absorption and Multiplication (SAM) structure in which the long wavelength radiation is absorbed in a narrow bandgap region (InGaAs), while the multiplication takes place in a wide bandgap layer (InP). The reasons for the SAM APD design are described elsewhere[3], and are not discussed here. While many InGaAs APD's are fabricated using a single epitaxial growth sequence for the required layers, and make use of a separately diffused guard-ring to avoid edge breakdown, the APD structure of this paper allows the diffused junction region to overlap the region of multiplication, as defined by the area of implantation, thereby avoiding high fields and breakdown in the peripheral region. The general field configurations in the two regions of the device are shown in Figure 2. While the simple one-dimensional model shown does not take into consideration the field enhancement due the curved edges of the diffused region, the field is still well below that required to cause multiplication. There are several advantages to the design:

1. Ion implantation is a reliable and reproducible method for introducing the required carriers in the multiplying region;

2. Edge breakdown problems are avoided by the simple expedient of allowing the diffused junction region to overlap the implanted area; and

3. The desired characteristics of the APD are somewhat less sensitive to fabrication parameters than designs which depend on the growth of a doped layer for the multiplication region.

A similar structure, but fabricated using different techniques, has recently been described by Tarof et al.[4]

3. FABRICATION

In this work, Vapour Phase Epitaxy was used to grown the various layers required. With reference to Figure 1, the thickness of the InGaAs light absorption region was 4 μm, in order to ensure good quantum efficiency at 1540 nm. The implantation is silicon, and net doses have varied between 2.2 and 2.6 x $10^{12}/cm^2$. The thickness of the InP regrowth layer was between 3.0 and 3.4 μm. Both InP and InGaAs layers are grown nominally undoped, and net concentrations are usually found to be less than about 2 x $10^{15}/cm^3$. Diffusion of the junction is at 450°C from a zinc source, and for the parameters listed above, diffusion times range between about 100 and 180 minutes.

In contrast to device requirements for telecommunications applications, where speed of response and high gain-bandwidth product are parameters of great importance, the rangefinding applications are relatively narrow band, typically less than 100 MHz. Thus, while high gain-bandwidth product APD's require very narrow multiplying regions[5], the devices for applications discussed here may have fairly wide multiplying regions.

In this work, some computer modelling has been carried out to calculate the interdependence of the various fabrication parameters. While those calculations are not presented here, a few general points are worth noting. For a desired range of operating conditions and gain-voltage characteristics, the following observations are made:

1. Wider multiplying regions are achieved with lower implant doses, which, for a given thickness InP capping layer thickness, means lower diffusion times;

2. Diffusion times or InP capping layer thicknesses are somewhat less critical for wider multiplying regions;

3. Breakdown voltages are higher for wide multiplying regions;

4. The peak field in the multiplying region is lower for wider depletion layers.

While intuitively most of these observations may be somewhat obvious, the calculations have been useful in demonstrating the considerable flexibility of the design and fabrication method. In this application, it is noted that there is some advantage to having devices with lower fields in the multiplying region in order to have improved excess noise factors.[6,7]

An evaluation of many devices fabricated using the techniques described above has shown that the best noise and dark currents are obtained for APD's having the lowest electric field in the InGaAs region. For devices fabricated in this work, the best range of maximum fields in the InGaAs is between about 0.5 and 1×10^5 volts/cm, which corresponds to voltages above "reach-through" between about 10 and 30 volts. Operation in this voltage and field range is possible since the bandwidth requirement is minimal, and operation at low gains, as is often the case for telecommunications applications, is not required.

4. PERFORMANCE CHARACTERISTICS

The relative response characteristic of an 85 μm diameter APD is shown in Figure 3. The curve has been normalized to unity for the response region just below the voltage at which the depletion layer reaches through into the InGaAs region (at about 38 volts). Thus, at all voltages except those below about 20 volts, the curve can be read as gain. Below about 38 volts, there is no depletion layer in the central (multiplying) region of the device, and carriers are collected by diffusing to the peripheral region, in which there is a depletion layer. Below 20 volts, there is no depletion layer in the InGaAs layer, even in the peripheral region, and therefore no collection of carriers. While gains as high as 80 or more, as shown, are usually possible, in a practical application, gains above about 20 are not likely to be useful due to the high noise. Figure 4 shows the dark current and noise for the diode, as a function of the gain. The measurements were made at a temperature of 23°C. It is noted that the total dark current at a gain of 10 is 32 nA, while the noise is less than 0.5 pA/$\sqrt{\text{Hz}}$.

The noise spectral density in an avalanche photodiode is given by:[8]

$$S = \frac{i_n^2}{B} = 2q\left[I_s + \left(I_b + I_{ph}\right) M^2 F\right] \tag{1}$$

where S is the noise spectral density (Amps^2/Hz); I_s is the surface dark current (not multiplied); I_b is the component of the dark current which undergoes multiplication; I_{ph} is the photocurrent induced in the multiplying region; M is the gain of the APD; F is the excess noise factor and B is the bandwidth. The excess noise factor is given by:

$$F = kM + (1-k)(2-1/M) \tag{2}$$

where k is the effective ratio of the ionization coefficients.

By making measurements of the illuminated noise in the APD, it is possible to determine the excess noise factor as a function of the gain, and derive from this the effective ratio of the ionization coefficients. This has been done in Figure 5. The input current level for the measurement was 100 nA, so that the effect of the surface and bulk dark currents could be neglected in the calculation of F from Equation (1), with minimal error. An approximate value for k is determined by fitting a calculated curve at M = 10. When this is done, a value for k of 0.38 is obtained. Reasonably good fit to the experimental data is obtained in the gain range between about 10 and 20. Above M = 20, some deviation occurs, possibly due to gain non-uniformity, while below M = 10, the statistics of the multiplication process in the very narrow multiplying region of these devices, result in lower noise than is predicted by McIntyre's theory.[8]

Noise and dark current are plotted for a 200 μm front-entry detector in Figure 6. Here, the curves have been plotted for responsivity at 1540 nm, rather than gain. While it was possible to achieve responsivities of at least 30 A/W in this device, a practical operating point might be about 10 A/W, for which the dark current was about 76 nA, and the noise about 0.75 pA/√Hz.

With the use of a spherical (ball) lens having a refractive index of about 1.9 and mounted on an APD, it is possible to increase the effective diameter of the device, depending somewhat on the diameter of the ball, and the f number of the incoming radiation. In this work, a ball lens of diameter 600 μm has been mounted on a back-entry version of the 200 μm APD. This extends the effective diameter of the diode up to at least 500 μm for applications in which the f number of the incident radiation is greater than about 1.5. A schematic diagram of the concept is shown in Figure 7. Dark current and noise obtained for the lensed APD and illuminated with a 500 μm spot (optics ~ f/1.8) are shown in Figure 8. For this device, the overall quantum efficiency was found to be about 65%, so that a responsivity at 1540 nm of 8 A/W in the figure corresponds to a gain of about 10. At this point, the dark current and noise of the detector were, respectively, 152 nA and 1.4 pA/√Hz. An optical scan of the APD, operating at a gain of approximately 10, is shown in Figure 9. The mixing effect of the ball lens is clearly evident in the excellent response uniformity obtained.

Capacitance of the 200 μm diameter detectors, both front and back entry, is found to be about 2 pF. The 3 dB frequency for these devices is typically greater than 400 MHz.

5. CONCLUSIONS

The design and performance characteristics of InGaAs APD's suitable for eye-safe rangefinding applications have been described. Although noise characteristics would probably limit the range of useful gains in most applications to less than about 20, well behaved performance has been demonstrated to gains as high as 80 in 85 μm diameter devices, and to at least about 30 for 200 μm diodes. The back-entry APD, with mounted ball lens, shows good characteristics for detection in a 500 μm diameter area.

While the largest APD's fabricated using the design and fabrication techniques discussed here have been the front and back entry versions of the 200 μm APD, it should be possible to make larger detectors to meet the requirements of more demanding systems. For example, a 300 μm diameter device used with the 600 μm ball lens should be suitable for achieving an effective diameter of at least 500 μm in a system having f/1 optics. Although there would be a noise, dark current and capacitance penalty to be paid for the increased area, it is believed that useful operation should be possible up to responsivities in the 7 to 10 A/W range at 1540 nm wavelength.

Operation of the APD's at lower temperatures is possible, to further reduce noise and dark current for M = 10, or to allow for operation at higher values of gain with acceptable noise characteristics. The activation energy for temperature dependence of the multiplied portion of the dark current is found to be approximately 0.4 eV. Thus, operation at -20°C will result in a reduction of the noise by nearly 4 times, or make possible an increase in gain from 10 to about 25, for the same noise.

6. REFERENCES

1. P. P. Webb, R. J. McIntyre, M. Holunga, and T. Vanderwel, "Planar InGaAs/InP APD Fabrication Using Silicon Implantation and Regrowth Techniques", Components for Fiber Optics Applications II, Ed. Vincent J. Tekippe, SPIE Proc. Vol. 839, pp.148-154, August, 1987.

2. P. P. Webb, R. J. McIntyre, J. Scheibling and M. Holunga, "Planar InGaAs/InP Avalanche Photodiode

Fabrication Using Vapour-Phase Epitaxy and Silicon Implantation Techniques", Technical Digest, Optical Fiber Communications Conference, New Orleans, p.129, 1988.

3. See, for example, K. Nishida, K. Taguchi and Y. Matsumoto, "InGaAsP heterostructure avalanche photodiodes with high avalanche gain", *Appl. Phys. Lett.*, Vol. 35, pp.251-252, 1979.

4. L. E. Tarof, D. G. Knight, K. E. Fox, C. J. Minar, N. Puetz and H. B. Kim, "Planar InP/InGaAs avalanche photodetectors with partial charge sheet in device periphery", *Appl. Phys. Lett* 57, p.670, 1990.

5. Mansanori Ito, Takashi Miawa and Osamu Wada, "Optimum design of δ-doped avalanche photodiode by using quasi-ionization rates", *J. Lightwave Technology*, 8, No. 7, p.1046, 1990.

6. L. W. Cook, G. E. Bulman and G. E. Stillman, "Electron and Hole Impact Ionization Coefficients in InP Determined by Photo-Multiplication Measurements", *Appl. Phys. Lett.*, Vol. 40, No. 7, pp.589-591, 1982.

7. Kenko Taguchi, Toshitaka Torikai, Yoshimasa Sugimoto, Kikuo Makita and Hisahiro Ishihara, "Temperature dependence of impact ionization coefficients in InP", *J. Appl. Phys.*, 59, pp.476-481, 1986.

8. R. J. McIntyre, "Multiplication noise in uniform avalanche diodes", *IEEE Trans. Electron Devices*, Vol. ED-13, pp.164-168, 1966.

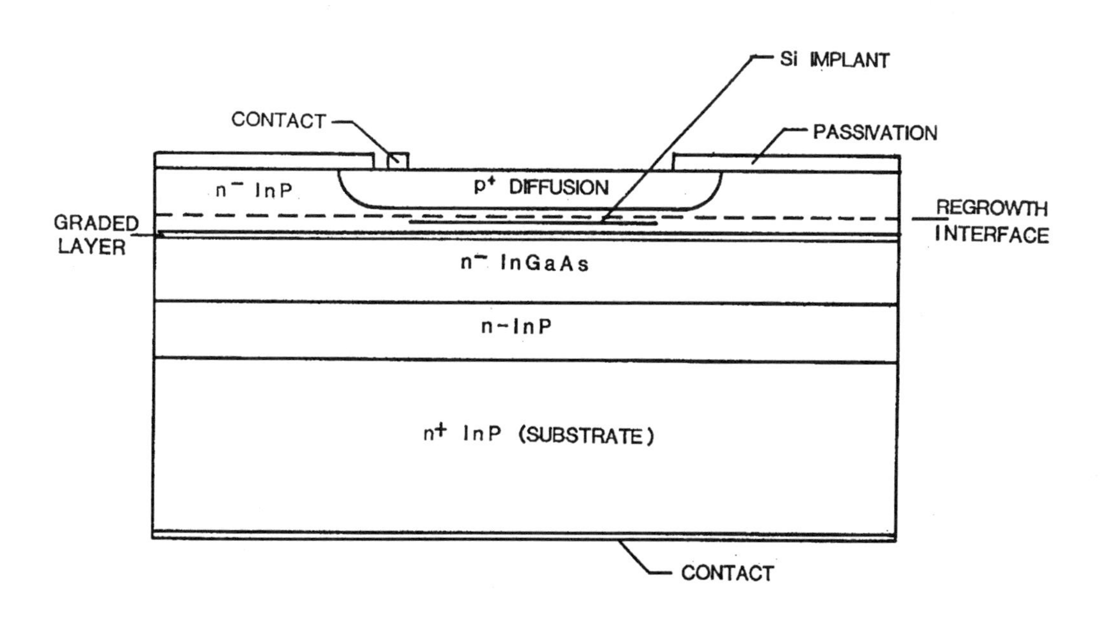

Figure 1 InGaAs/InP APD structure, using silicon implanted multiplying region.

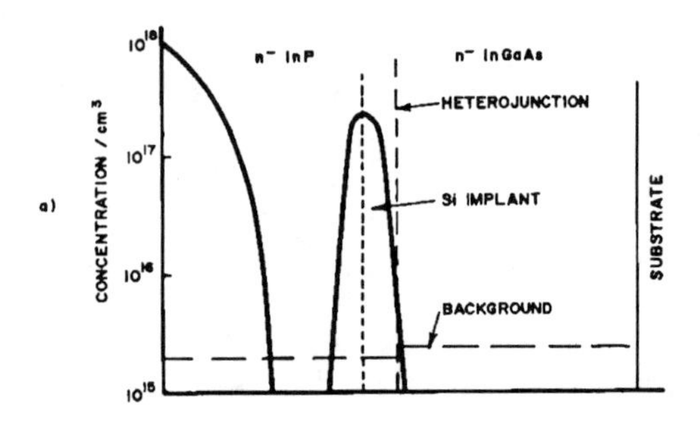

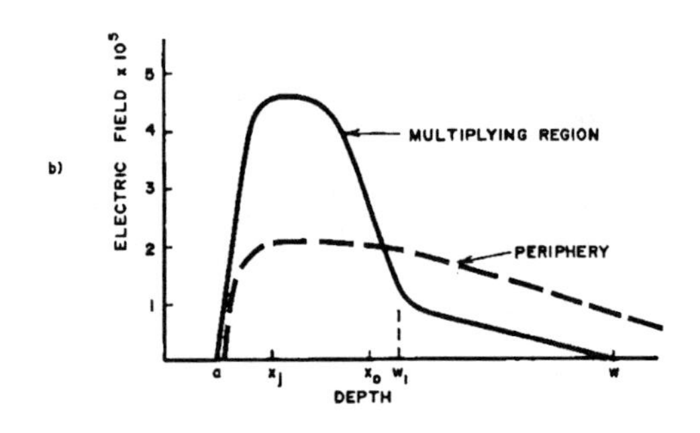

Figure 2 (a) Impurity distribution of implanted APD structure; (b) Field profile.

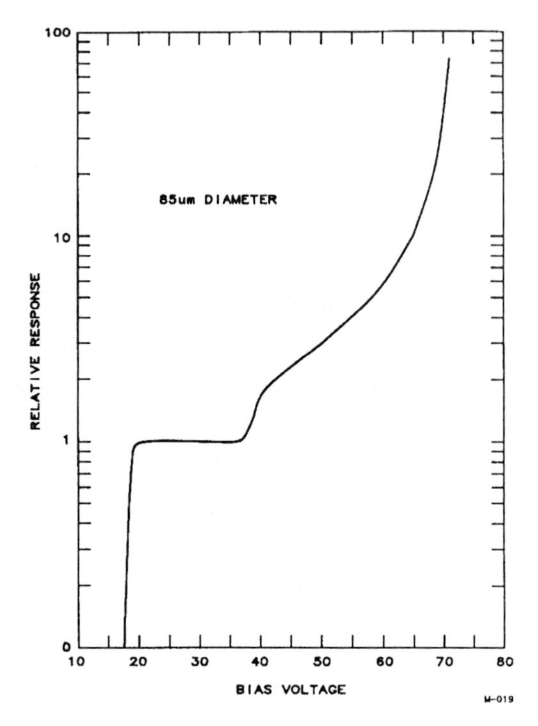

Figure 3 Relative response - voltage curve; 85 μm dia. APD

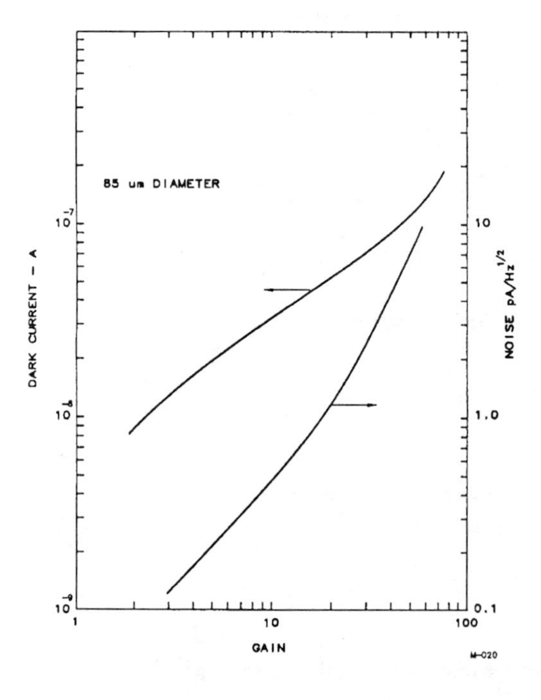

Figure 4 Dark current and noise vs. gain; 85 μm dia. APD.

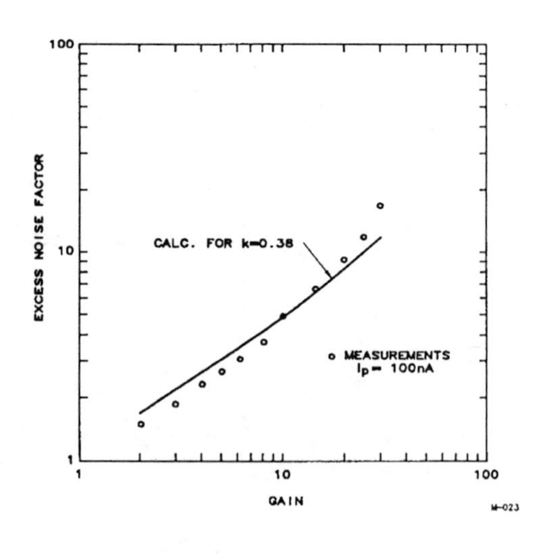

Figure 5 Excess noise factor vs. gain; 85 μm dia. APD.

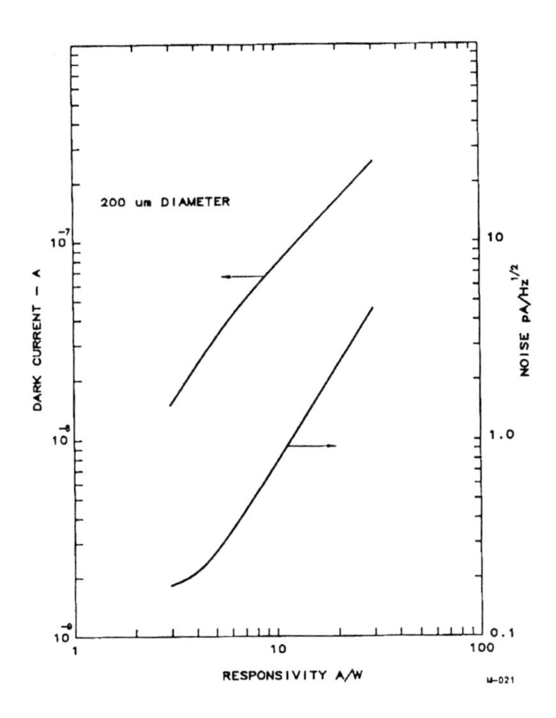

Figure 6 Dark current and noise vs. responsivity; 200 μm dia. APD

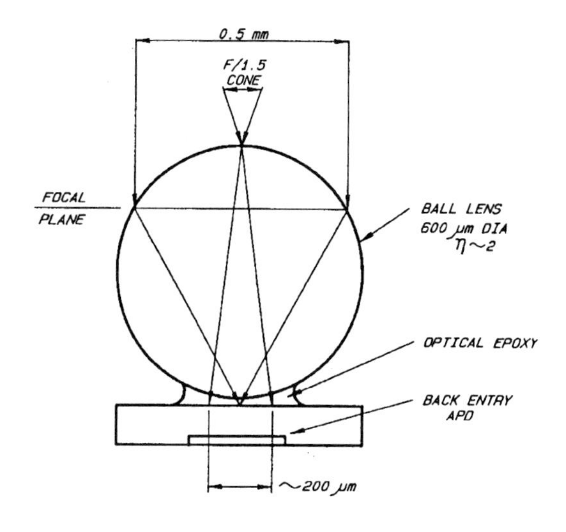

Figure 7 Schematic diagram, lensed back-entry APD

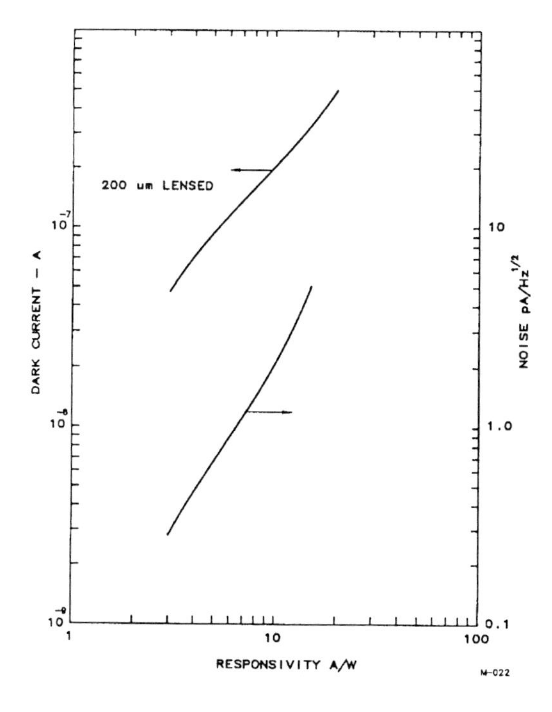

Figure 8 Dark current and noise vs. responsivity; lensed APD

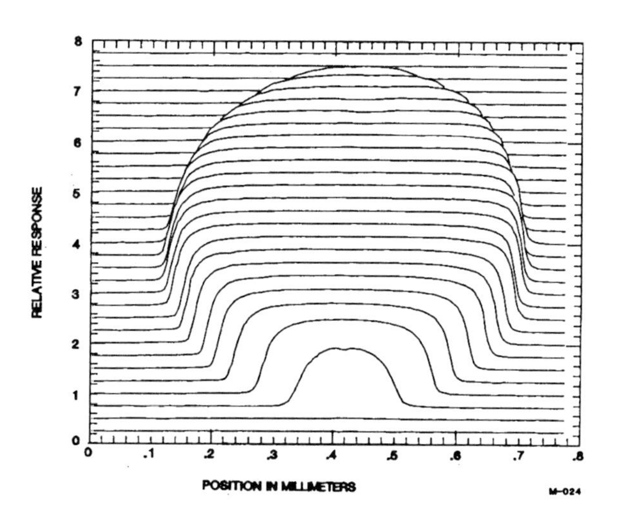

Figure 9 Response uniformity, lensed APD

High Performance InGaAs PIN and APD Detectors for 1.54 um Eyesafe Rangefinding

G.H. Olsen, D.A. Ackley, J. Hladky, J. Spadafora, K. Woodruff, M. Lange and B. VanOrsdel

EPITAXX, Inc., 3490 U.S. Route One, Princeton, NJ 08540

and

S.R. Forrest and Y. Liu, University of Southern California, Los Angeles, CA 90089

Abstract

The structure and device properties of indium gallium arsenide (InGaAs) pin photodiodes and avalanche photodiodes (APDs) are described. Quantum efficiencies above 85% at 1.54 um, dark current densities near 1 uA/cm^2 (-5V, 300K) and 3 mm diameter shunt resistances (10 mV, 300K) above 10 megohms have been observed. Avalanche gains above 20 have been measured with multiplied primary dark currents below 7 nA. Extended wavelength $In_xGa_{1-x}As$ (.53 < x < .80) pin detectors are also described with 70% quantum efficiency and room temperature RoA products above 2000 ohm -cm^2 at 1.8 um, 900 ohm -cm^2 at 2.1 um and 15 ohm -cm^2 at 2.6 um.

Indium Gallium Arsenide pin Detectors

The direct bandgap structure of InGaAs offers much lower dark current, higher quantum efficiency, and lower noise equivalent power than those of indirect bandgap germanium (Ge). Typical InGaAs quantum efficiency at 1.3 um is 85% (vs. 50 to 60% for Ge) while the InGaAs noise properties are an order of magnitude better than those of Ge. Room-temperature InGaAs will outperform[(1)] cooled Ge.

$In_{0.53}Ga_{0.47}As$ is the alloy with the smallest bandgap (Eg) that can still achieve lattice matching with indium phosphide. The lowest wavelength that any substance can absorb is set by the inverse of the bandgap energy (1.24/Eg). Although the small bandgap (0.75 eV) of $In_{0.53}Ga_{0.47}As$ results in larger dark currents than those of quaternary $In_{.7}Ga_{.3}As_{.5}P_{.5}$ with a cutoff wavelength just beyond 1.3 um, it absorbs light out to 1.7 um and still has considerably lower dark current than that of germanium.

InGaAs detector technology[(2)] has borrowed heavily from modern silicon technology. Planar technology evolved from the older mesa technology, silicon nitride (SiN_x) has replaced silicon dioxide and organic materials for passivation, and eutectic die bonding has replaced epoxy mounting to give improved performance and vastly superior quality. A typical planar InGaAs detector strucuture is shown in Figure 1. The hydride vapor phase epitaxy (VPE) technique[(2)] produces high quality layers of InGaAs and InP with the capacity of over six 2" diameter wafers per day.

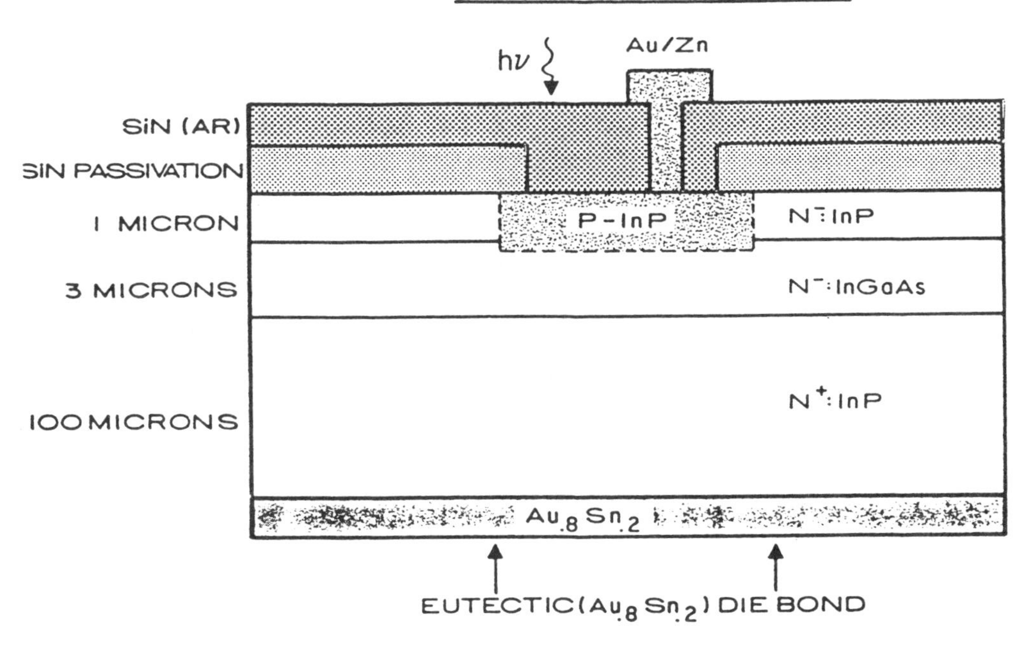

Figure 1. Cross-section of planar InGaAs/InP detectors.

The p/n junction is formed by a zinc diffusion through a hole opened in the SiN. The periphery of the p/n junction is thus completely protected by the SiN, and, furthermore, is terminated in a high-bandgap (i.e., lower leakage current) InP layer which also serves to decrease surface recombination (and thus increase quantum efficiency by ~20-30%) at the top InGaAs surface.

Another advantage of the planar structure shown in Figure 2 is the thin (~1 um) InP top layer. Although InP absorbs strongly below 0.9 um, thin InP can allow appreciable visible light to pass through and quantum efficiencies above 30% have been measured at 0.82 um with planar InGaAs detectors. Such response allows one InGaAs detector to replace <u>both</u> an Si and Ge detector, especially in applications such as optical power meters. Quantum efficiencies near 50% at 0.5 um have been measured with InGaAs detectors having InP caps thinned below 0.5 um.

The silicon nitride passivation, planar structure, and eutectic $Au_{0.8}Sn_{0.2}$ die bonding employed with planar VPE InGaAs detectors allow burn-in and reliability studies to be performed at temperatures beyond 200C and -20V reverse bias. Such studies predict room temperature operating lifetimes of over 10^{11} hours or a failure rate of < 0.01 FIT. Reliabilities comparable to those of silicon integrated cicuits have already been demonstrated for VPE planar InGaAs detectors.

Table 1 contains a summary of typical and best performance characteristics of standard $In_{.53}Ga_{.47}As$ pin detectors. More than an order of magnitude decreases in dark current has resulted in the last two years from evolutionary process improvements.

InGaAs Detectors Results

Diameter (um)	Typical QE	Typical I_d	Typical R_{sh}	Best QE	Best I_d	Best R_{sh}
ETX-300	85	1		95	0.1	
ETX-500	85		100	95		150
ETX-1000	85		35	95		100
ETX-2000	85		10	95		50
ETX-3000	85		2	95		30

I_d = nA @ -5V

R_{sh} = megohms @ ±5 mV

QE = % @ 1300 nm

InGaAs Avalanche Photodiodes (APDs)

$In_{0.53}Ga_{0.47}As$/InP avalanche photodiodes (APDs) are of substantial interest for eyesafe rangefinding applications at 1.54 um that require high detector sensitivity. Extensive strides have been made in fabricating[4] reliable, planar APD structures, but relatively complex device structures have been required to control the edge gain. InGaAs APDs offer high gain at 1.54 um and thus can substantially increase eyesafe rangefinding system performance.

Figure 2 shows a separate absorption and multiplication region (SAM-APD) structure with a simple double-diffused structure that effectively suppresses edge gain. The device combines a floating guard ring design[5], previously used in both Si power devices and InP/InGaAs SAM APDs with additional guarding introduced by a second diffusion[6] and grading of the p-n junctions at the edges of the active area of the device. While two-component diffusions have been applied to planar guard ring APD structures before, the guard rings in those devices were formed by long (300 h) initial Cd diffusions, followed by a deeper Zn diffusion. This is in contrast to the process described here which simply utilizes the fixed-source diffusion of Zn under a SiN_x passivation layer to grade the doping profile. The junction grading, combined with the floating guard rings, results in an extremely robust design with uniform gains as high as 85, dark currents at 90% of breakdown below 5 nA, and low capacitance.

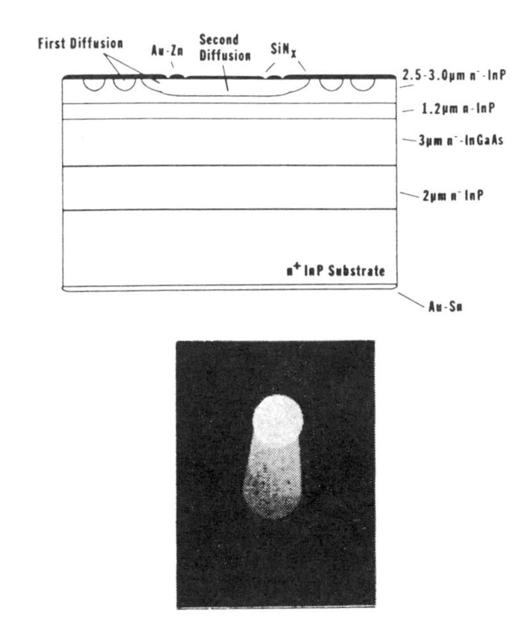

Figure 2. Schematic cross section and top view of a completed front-side illuminated floating guard ring APD fabricated by double diffusion.

The detector is front-side illuminated with a 75 um diameter active area and a 50 um diameter circular bonding pad. The floating rings, which completely surround the active area, are readily seen in the photograph. The tear-drop shape is utilized to minimize the additional capacitance introduced by the bonding pad area. One feature of the floating ring design is that the rings do not noticeably increase the device capacitance because they are effectively connected in series with the active layer.

A typical layer structure consists of a 2 um InP buffer ($n \sim 2 \times 10^{16} cm^{-3}$), a 3 um thick InGaAs absorbing layer ($n < 2 \times 10^{15} cm^{-3}$), an InP multiplication layer with a thickness of 1.25 um and doping of $2.1 \times 10^{16} cm^{-3}$, and a 2.5-3.0 um undoped InP cap with a background doping of $1.5 \times 10^{15} cm^{-3}$. Device fabrication begins with a first Zn diffusion through a SiN_x mask photolithographically patterned in the floating ring configuration. A second SiN_x layer is then deposited, and a window whose radius is 8-10 um less than the initial diffusion in the central p-n junction region is aligned to each device. The alignment between the first and second diffusions is to within 2-3 um. During the second diffusion, the region under the window is diffused to within approximately 0.5 um of the mulitplication layer. In addition, the Zn-diffused regions still under the SiN_x also diffuse in deeper, but they are effectively diffusing under constant source conditions which results in grading of the p-doping profile. Secondary ion mass spectrometry (SIMS) data show that the logarithmic slope of the p-dopant profile is reduced by a factor of three at the junction level, substantially increasing the guarding at the junction periphery. The small differences in the diffusion rates in the two regions make it straightforward to achieve a depth separation between the two fronts of 0.5 - 1.0 um, which is optimal for guarding the structure.

Excellent characteristics have been observed for devices fabricated by the above process. A plot of the dark and photo currents and gain as a function of reverse bias for a typical device is shown in Figure 3. The breakdown voltage is about 76V, and were typically in the range 70-85V for most of the devices on the wafer. The dark current at 90% of breakdown ($0.9V_b$) is 11 nA, which corresponds to a primary dark current of < 2nA. In the very best devices, dark currents as low as 0.5 nA at 0.9 V_b have been observed. The maximum gain for the device in Figure 3 is > 25 which is typical of high-quality devices. The best devices had gains as high as 85 at 100 nA dark current. Quantum efficiencies for devices without antireflection coatings have been estimated to be in excess of 70%. The device capacitance has been determined to be as low as 0.4 pF at 90% of breakdown. Thus, the device appears to be suitable for high speed, high sensitivity applications at 1.54 um.

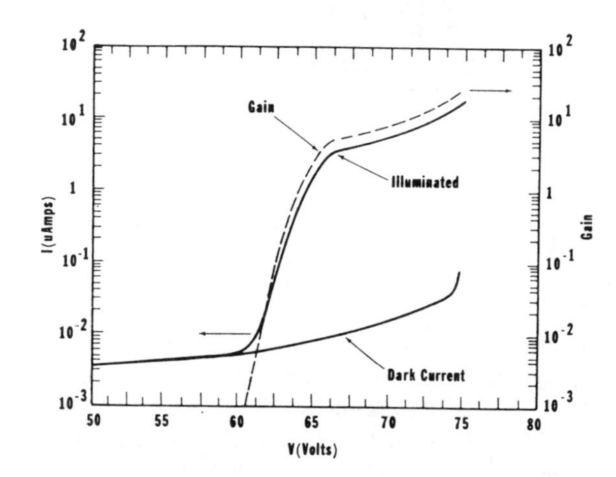

Figure 3. I-V and gain characteristics of a floating guard ring APD.

Extended Wavelength Detectors

Finally, eyesafe rangefinding can also be performed near 2 um due to the availability of solid state lasers in this wavelength range. Detectors for this range have been described previously(7) and the structure for an $In_{.8}Ga_{.2}As$ device which absorbs out to 2.6 um is shown in figure 4. The high growth rate (~25 um/hr) and materials flexibility of the hydride VPE crystal growth technique are particularly suited to the fabrication of these structures.

Compositionally Graded VPE $In_{.82}Ga_{.18}As/InAs_{.6}P_{.4}$ 2.6µm Planar Detector

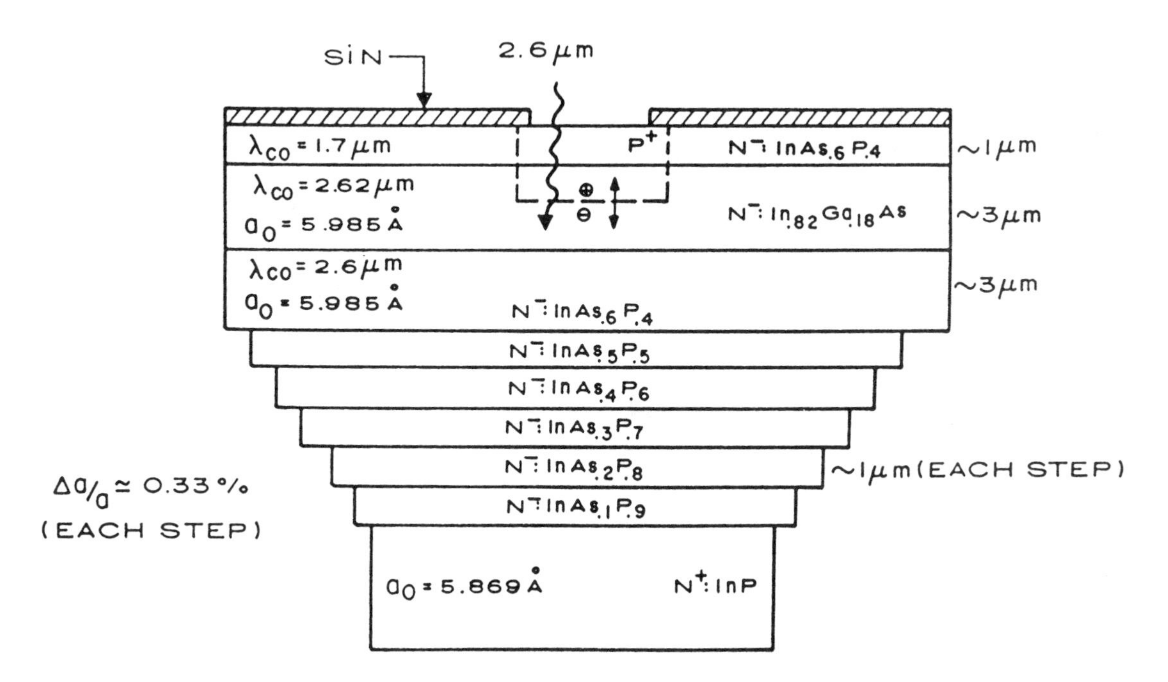

Figure 4. Compositionally graded detector structure.

Data from wafers with 1.8 (X = 0.6), 2.1 (X = 0.7) and 2.6 um cutoffs are shown in table 2. Detectors with 75, 100, 300, 500 and 1000 um diameters were made on the same wafer. All data shown was taken at room temperature. Applications for these devices include spectroscopy, LIDAR and windshear detection.

Properties of Extended Wavelength InGaAs Detectors

	Diameter (um): 75	100	300	500	1000
1.8 um					
I_d (nA @ -1V)	2	9	73	198	750
J (uA/cm^2)	48	115	103	101	95
R_{sh}(Mohm)	67	20	2	.83	.31
QE (% @ 1.75)	75	75	75	75	75
R_0A(Ω-cm^2)	3000	1570	1413	1629	2434
2.1 um					
I_d (nA @ -1V)	30	10	100	180	1790
J (uA/cm^2)	679	127	141	91	228
R_{sh} (Mohm)	20	12	1.1	.50	.13
QE (% @ 1.9)	70	70	70	70	70
R_0A (Ω-cm^2)	883	942	777	981	1021
2.6 um					
I_d (nA @ -1V)	321	650	4389	10678	34000
J (uA/cm^2)	7287	8276	6211	5441	4337
R_{sh} (Mohm)	0.22	.13	0.015	0.005	0.002
QE (% @ 2.5)	50	50	50	50	50
R_0A (Ω-cm$_2$)	10	11	11	10	16

References

1. G.H. Olsen, "Why InGaAs Detectors Outperform Germanium", Lightwave, 2, 31-34 (1986).

2. G.H. Olsen and V.S. Ban, "InGaAsP: The Next Generation in Photonics Materials", Solid State Technology, 30, 99-106 (1987).

3. S.R. Forrest, V.S. Ban, G. Gasparian, D. Gay and G.H. Olsen, "Reliability of Vapor Grown Planar InGaAs Photodiodes", IEEE Electron Dev. Lett., 9, 217-219 (1988).

4. P. Webb, R.J. McIntyre, J. Scheibling, and M. Holunga, "Planar InGaAs/InP Avalanche Photodiode Fabrication Using Vapor-Phase Epitaxy and Silicon Implantation Techniques", in Proc. Opt. Fiber Communn. Conf. Tech. Series, Vol. 1, pp. 129-130 (1988).

5. Y. Liu, S.R. Forrest, V.S. Ban, K.M. Woodruff, J. Colosi, G. Erickson, M.J. Lange and G.H. Olsen, " A Simple Very Low Dark Current, Planar, Long Wavelength Avalanche Photodiode", Appl. Phys. Lett. 53, 1311-13 (1988).

6. D.G. Ackley, J. Hladky, M.J. Lange, S. Mason, G. Erickson, G.H. Olsen, V.S. Ban, Y. Liu and S.R. Forrest, "$In_{0.53}Ga_{0.47}As$/InP Floating Guard Ring Avalanche Photodiodes Fabricated by Double Diffusion", IEEE Photonics Technology Letters, 2, 571 (1990).

7. G.H. Olsen, A.M. Joshi, S.M. Mason, K.M. Woodruff, E. Mykietyn, V.S. Ban, M.J. Lange, J. Hladky, G.C. Erickson and G.A. Gasparian, "Room-Temperature InGaAs Detector Arrays for 2.5 um", Proceedings of SPIE, Vol. 1157, Orlando (1989).

8. G.H. Olsen, S. Mason, G. Gasparian, A. Joshi, K. Woodruff, K. Linga, D. Rodefeld, M. Lange, F. Speer, V. Ban and G. Erickson, "Room-Temperature Properties of InGaAs Detectors Optimized for 1.8, 2.1 and 2.5 um", Proc. IEEE Lasers and Electro Optics Society (LEOS) Conference, (Boston, Nov., 1990).

Testing laser eye protection

Jack A. Labo
Michael W. Mayo

Armstrong Laboratory
Directed Energy Division
Brooks Air Force Base, Texas 78235-5301

ABSTRACT

All laser eye protection (LEP) should be tested to insure that it meets certain performance criteria. There are no American standard test methodologies other than military standards for LEP to meet performance criteria. Depending on the type of eyewear, the actual protection afforded during laser exposure may be different from the protection measured by standard methods. Specific user application environments may require additional tests for laser eyewear such as solar, temperature, and humidity stability. Optical quality testing methodologies including optical distortion, refractive power, prismatic deviation, haze, and chromaticity coordinate measurements are discussed along with the importance of user acceptance in optical quality, field-of-view, weight, comfort, corrective lens compatibility, and style.

1. INTRODUCTION

Since the laser was invented, the need for eye protection from the effects of coherent radiation for workers performing duties in the vicinity of laser systems has been of utmost concern[1]. Over the last 30 years, the quality and integrity of laser eye protection (LEP) has considerably improved. LEP from both narrow and broad spectral bands is commercially available in a variety of configurations.

The assessment of early eye protection revealed serious limitations and technical problems such as optical bleaching, poor stability, low damage thresholds, and poor vision[2]. Most of the eye protection was cumbersome, unattractive, and incompatible with corrective lenses. Many improvements have been made over the years to create eyewear which is lightweight, comfortable, technically stable, and even stylish! However, there are still limitations, and that ideal material which protects against all laser hazards remains elusive. Therefore, one must keep in mind that LEP should not solely be relied upon to protect against all laser hazards; LEP should not be a substitute for engineering controls or appropriate laser operating practices. A publication such as the *Guide for the Selection of Laser Eye Protection* is useful to assess eye hazards and match protection to the particular hazard[3].

Acceptable eye protection must (1) provide enough protection at the wavelengths which pose a threat, (2) have adequate photopic and scotopic luminous transmittance for a given task, (3) have excellent optical quality, (4) be lightweight for comfortable prolonged use, and (5) be able to be worn with spectacles or other required protective gear. The following areas should also be considered in selecting suitable protective eyewear: other laser output characteristics such as pulsewidth, pulse repetition frequency, energy or power, spot size, and beam divergence; damage threshold and photosaturation characteristics of the material; field-of-view of the device; impact resistance requirements; operating environment; and overall comfort and fit.

2. OPTICAL PROTECTION CRITERIA

From ANSI Standard Z136.1-1986, the enclosure of the ocular hazard zone is the preferred control method[4]. Laser protective equipment should only be used when other control measures for adequate protection against radiation levels above the MPE cannot be provided. When eye protection must be used, it should attenuate the irradiance or radiant exposure by a factor sufficient to reduce transmitted radiation to levels at or below the Maximum Permissible Exposure (MPE) for the possible viewing conditions. The MPE is defined as the level of laser radiation to which a person may be exposed without hazardous effects or adverse biological changes in the eye[4]. This critical attenuation factor, known as the optical density (D_λ), is defined as the negative logarithm to the base ten of the optical transmittance at a specific wavelength (T_λ):

$$D_\lambda = -\log_{10}(T_\lambda) \tag{1}$$

The minimum D_λ needed to reduce the potential eye exposure (H_p) to the MPE level is determined by

$$D_\lambda = \log_{10}(H_p/MPE) \tag{2}$$

The D_λ of protective materials can then be determined by measuring the transmittance at the wavelength(s) of interest, and then using Equation 1. Transmittance is most often measured using a standard spectrophotometer operating in the spectral region of concern (Figure 1). Spectrophotometers can measure optical densities up to 5.0 for a broad spectral region, as shown by the plot in Figure 2.

However, there are two limitations in using a spectrophotometer as the only method to assess the spectral transmittance of a protective device. The first, and most obvious limitation, is the spectrophotometer's inability to measure optical densities in LEP above 5.0. Certain methods have been employed to determine the optical density above this limit; such as mathematical formulae (Beer's Law of absorption) or extrapolated data. The D_λ at specific laser wavelengths can be measured using a laser densitometer system. Figure 4 is an overview of a CW laser densitometer. Design sources for both a continuous wave and pulsed densitometer can be found in the References[5,6].

The other limitation, considered to be more serious, is the fact that the spectrophotometer data test only the ability of the material to absorb extremely low flux densities of incoherent radiation. The spectrophotometer does not measure nonlinear optical transmission characteristics of the absorbing material. This laser saturation process in an absorber is characterized by a nonlinear optical transmission versus irradiance as shown in Figure 3. Previous studies have shown that certain laser absorbing dyes and glass used in protective devices initially absorb the incident laser radiation, then temporarily change properties and allow the material to transmit higher than expected energy levels[7,8]. This is not a new phenomena and many of the mechanisms involved with the bleaching problem are explained in a paper by Guillano and Hess in 1967[9].

These nonlinear transmission effects, commonly referred to as laser bleaching, reverse saturable absorption, self-induced transparency, reverse photosaturation, or laser induced change in transmission, may occur when a laser absorbing material is exposed to high peak irradiances. In other words, the laser beam itself can "bleach" some protectors, thus increasing the optical transmission at the wavelengths which are supposed to be attenuated. This is a potentially dangerous scenario, since false security can be more dangerous than not wearing eye protection.

Therefore, LEP should be evaluated under the worst case conditions. Since most LEP manufacturers do not test their materials for nonlinear transmission effects and since current U.S. safety standards do not directly address these issues, it is up to the end user to evaluate the ability of the LEP to block the radiation from these high peak power laser systems. A simple test can be performed by placing the eyewear at a location where the laser radiation hazard to the eye is greatest and measuring the transmitted radiant exposure or irradiance. If the detected radiant exposure or irradiance is above the MPE, then the eyewear should not be used for protection in that instance. On the other hand, eyewear which continues to absorb laser radiation levels above the MPE, and is not damaged should be considered acceptable. LEP manufacturers can be contacted for specific optical specifications, or data for some eyewear can be obtained from the literature[3,10].

Additionally, adequate protection should be provided for all possible viewing angles which could cause a potential eye injury particularly for dielectric stack reflective-type filters which have a high degree of angular variability.

3. OPTICAL EVALUATION CRITERIA

In selecting suitable LEP, one must consider characteristics other than the ability to attenuate the laser beam. These are important considerations because adequate protection should not just prevent ocular injury, but should also pass sufficient light to perform all work related tasks. Since much of the information required to perform adequately and safely in a laser environment is obtained through the visual system, it is desired that laser eyewear manufacturers meet the ANSI standard Z87.1-1989[11]. Poor vision through LEP can increase the probability of injury from electrical shock, tripping, or exposure to dangerous chemicals.

Poor lens quality can cause severe eye strain, tearing, headaches, and other discomfort. If a particular LEP exhibits any of these undesirable qualities, workers may not wear the device or only wear it for short periods. Therefore, we feel that the manufacturer should certify that their product exhibits adequate optical quality. Laboratory LEP should meet the ANSI standards on prismatic power, refractive power, astigmatism, distortion, prism imbalance, and haze. The standards and test methodologies are depicted in ANSI Standard Z87.1-1989[11].

Luminous transmittance and eyewear field-of-view (FOV) requirements would most likely be application-specific. If possible, the photopic luminous transmittance (PLT) should be high enough not to seriously degrade visual acuity or contrast sensitivity. The FOV should be sufficiently wide; it should not restrict vision and consequently movement or operation in the specific work environment. Since the ultraviolet (UV) spectral region is not needed for vision, the UV transmittance should be minimal to protect the user from UV hazards. The PLT (T_L) is usually specified by the manufacturer and is normally expressed as

$$T_L = \frac{\int_{380}^{780} T(\lambda)\bar{y}(\lambda)S(\lambda)d(\lambda)}{\int_{380}^{780} \bar{y}(\lambda)S(\lambda)d(\lambda)} \qquad (3)$$

where $T(\lambda)$ is the spectral transmittance, $\bar{y}(\lambda)$ is the relative luminous efficiency function, and $S(\lambda)$ is the relative spectral emittance of CIE light source Illuminant A with respect to the CIE Standard Colorimetric Observer[11].

The user should first test the eyewear by wearing it in the work area with all instrumentation on and the laser hazard completely blocked. A thorough check should be made to assure that the necessary visual field reveals such things as instrument status and warning lights, chemical hazards, and trip hazards. Any image distortions, blurring, or prismatic effects denotes an optical quality defect in the eyewear. The user should also observe very carefully anything viewed through the eyewear which becomes invisible because its color (wavelength) is blocked by the eyewear. Likewise, depending on the spectral transmittance of the eyewear, certain color cues may spectrally shift and be perceived as a different color after being transmitted through the eyewear.

Once aware of such color discrimination problems, one can work around the problem and change procedures. It is definitely not advisable to take off the eyewear or peer around the lenses to pick up visual cues while the laser is operating since the exact location of the laser beam at that instant may not be known.

Aside from the ANSI Z87.1-1989 standard mentioned, we consider that all commercially available eyewear should be capable of maintaining original specifications after long-term storage. They should also be required to maintain their protection after long term exposure to normal lighting conditions. Improvements in dye stability and the addition of UV inhibitors have made this a relatively trivial task for most current eyewear. However, we still consider that data relating to shelf life and spectral integrity should be available to those who purchase LEP[10,12].

There are some other special considerations to remember. To maintain good optical quality, all eyewear should meet abrasion resistance standards[12]. For special applications, eyewear may have to be tested to withstand harsh environmental conditions such as heat, humidity, and special hardening. Figure 5 depicts a San Antonio, Texas Fall, clear sky solar spectral irradiance. Figure 6 illustrates how this spectral irradiance might be duplicated with a solar simulator as was done recently to test the lifetime of LEP exposed to extreme sunlight[12].

4. CONCLUSIONS

MPE requirements, utility, user acceptance, and testing all play a role in finding suitable laser eye protection. The user must understand the actual and associated laser hazards in the work environment which are fundamental to insuring adequate LEP. Periodic visual inspection of LEP wear-and-tear and/or damage is also necessary.

5. REFERENCES

1. R. G. Allen, J. A. Labo, and M. W. Mayo, *Laser Eye Protection*, Society of Photo Optical Instrumentation Engineers Proceedings, Vol. 1207, Laser Safety, Eyesafe Laser Systems, and Laser Eye Protection (1990), pp 34-45.

2. D. R. Marston, P. C. Laudieri, and P. D. Walker, *Evaluation of Laser Eye Protectors Commercially Available*, SAM-TR-72-8, USAF School of Aerospace Medicine, Brooks Air Force Base, Texas (AD 746 293), 1972.

3. D. H. Sliney, H. D. Edmunds, J. F. Smith, J. A. Labo, and M. L. Wolbarsht, *Guide for the Selection of Laser Eye Protection*, Laser Institute of America (LIA), 7th Printing, Oct. 1989, Orlando.

4. American National Standards Institute, *Safe use of Lasers*, ANSI Z136.1-1986, New York.

5. B. P. Edmonds and M. D. Turner, *A Laboratory Laser Densitometer System*, SAM-TR-87-42, USAF School of Aerospace Medicine, Brooks Air Force Base, Texas (August 1988). Distribution authorized to DoD Components only; software documentation; 7 December 1987. Other requests for this document shall be referred to USAFSAM/TSKS (STINFO Officer).

6. J. Taboada and W. J. Fodor, *Pulsed Dye Laser Densitometry Using an Optical Delay*, Applied Optics, 16:1132, (May 1977).

7. T. L. Lyon and W. J. Marshall, *Nonlinear Properties of Optical Filters - Implications for Laser Safety*, Health Physics Vol. 51, No. 1 (July 1986)

8. A. H. Blumenthal and J. J. Mikula, *Evaluation of Air Force Laser Protective Devices*, U.S. Army Frankford Arsenal Report R-2098 (Nov 1973).

9. G. R. Giuliano and L. D. Hess, *Nonlinear Absorption of Light: Optical Saturation of Electronic Transitions in Organic Molecules with High Intensity Laser Radiation*, IEEE Journal of Quantum Electronics (Aug 1967).

10. K. R. Envall and R. Murray, *Evaluation of Commercially Available Laser Protective Eyewear*, Bureau of Radiological Health, HEW Publication (FDA) 79-8086 (May 1979).

11. American National Standards Institute, *Practice for Occupational and Educational Eye and Face Protection*, ANSI Z87.1-1989, New York.

12. R. G. Allen, M. D. Turner, S. L. Ramsey, J. A. Labo, and M. W. Mayo, *Quality Test and Evaluation of USAF Laser Eye Protection Visors*, USAFSAM-TP-89-21, USAF School of Aerospace Medicine, Brooks Air Force Base, Texas, July 1990. Distribution authorized to U.S. Government agencies only; proprietary information; 6 March 1990. Other requests for this document must be referred to USAFSAM/TSKS (STINFO Officer).

Notice: This paper does not contain any sensitive information from referenced limited distribution publications or presentations.

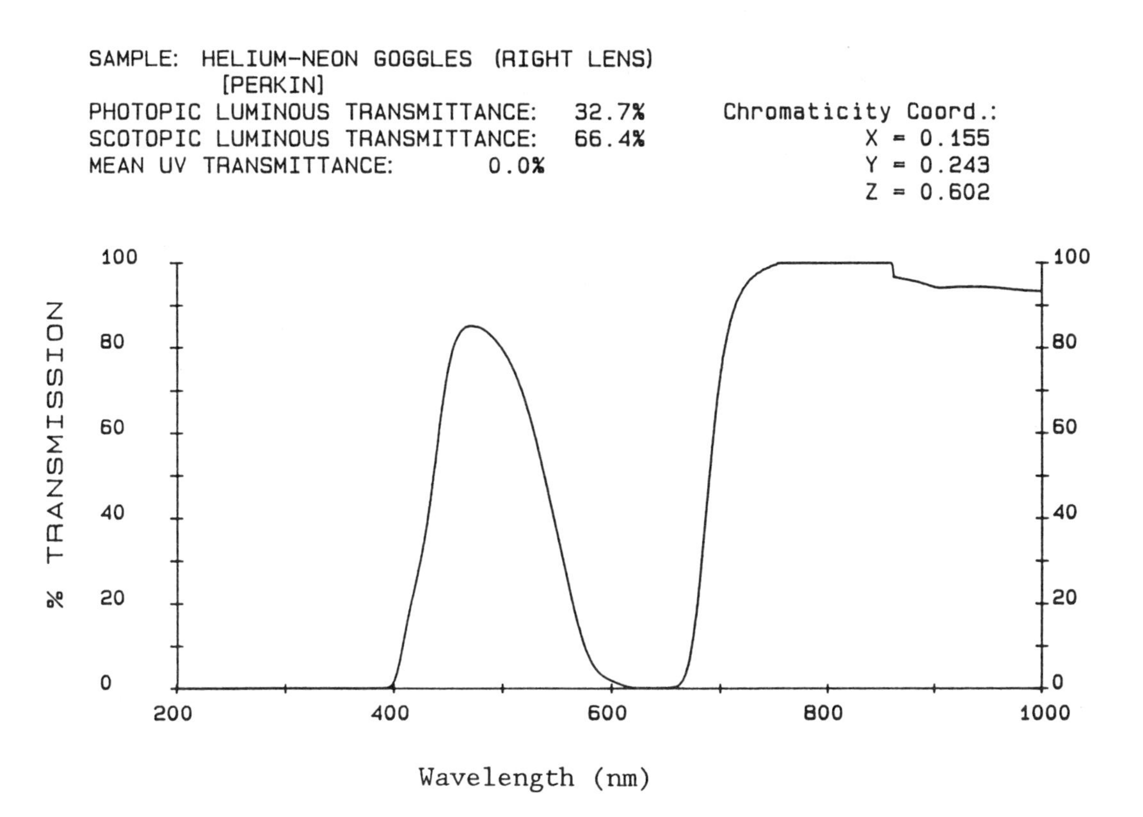

Figure 1. Spectral Transmittance.

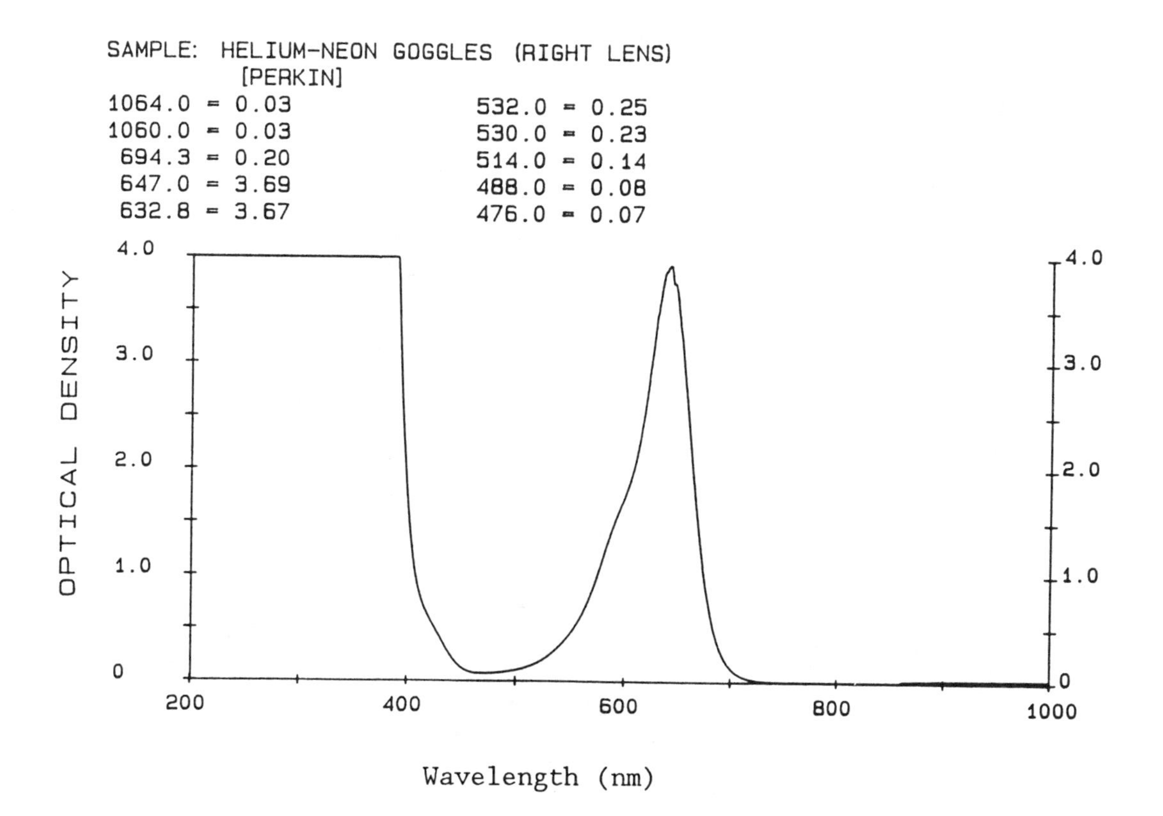

Figure 2. Spectral Optical Density

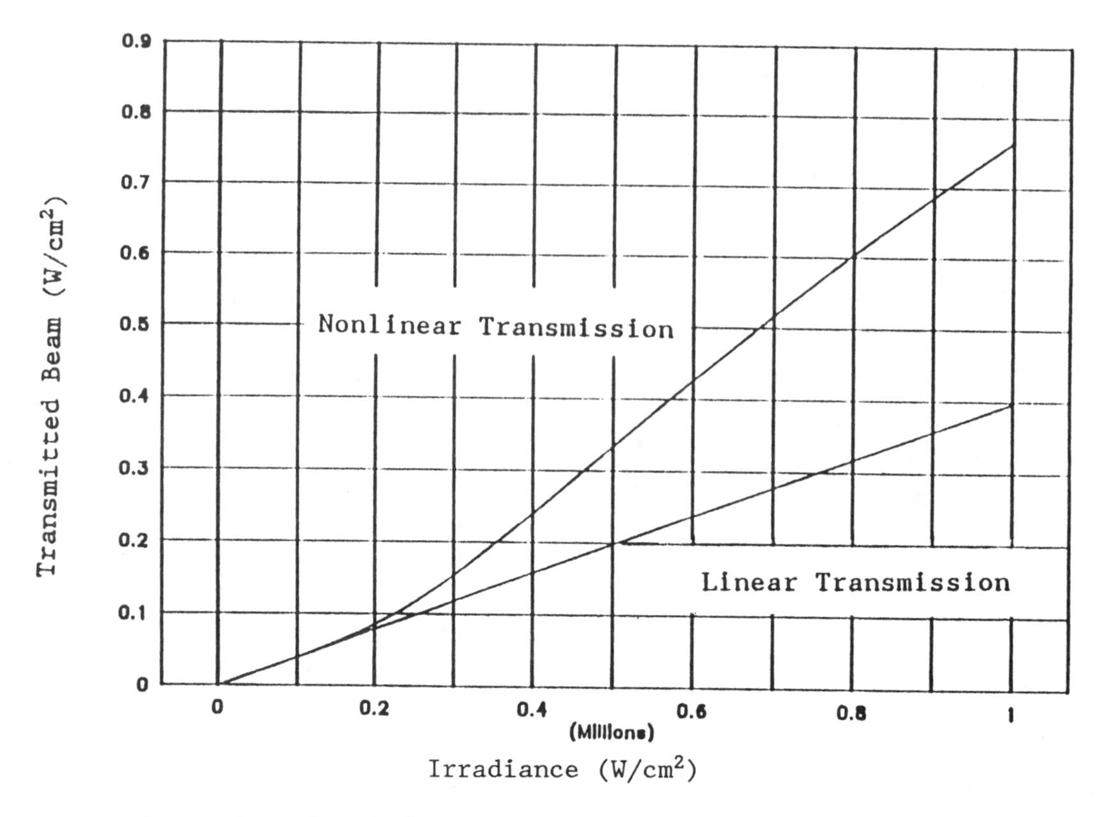

Figure 3. Threshold for Nonlinear Optical Transmittance.

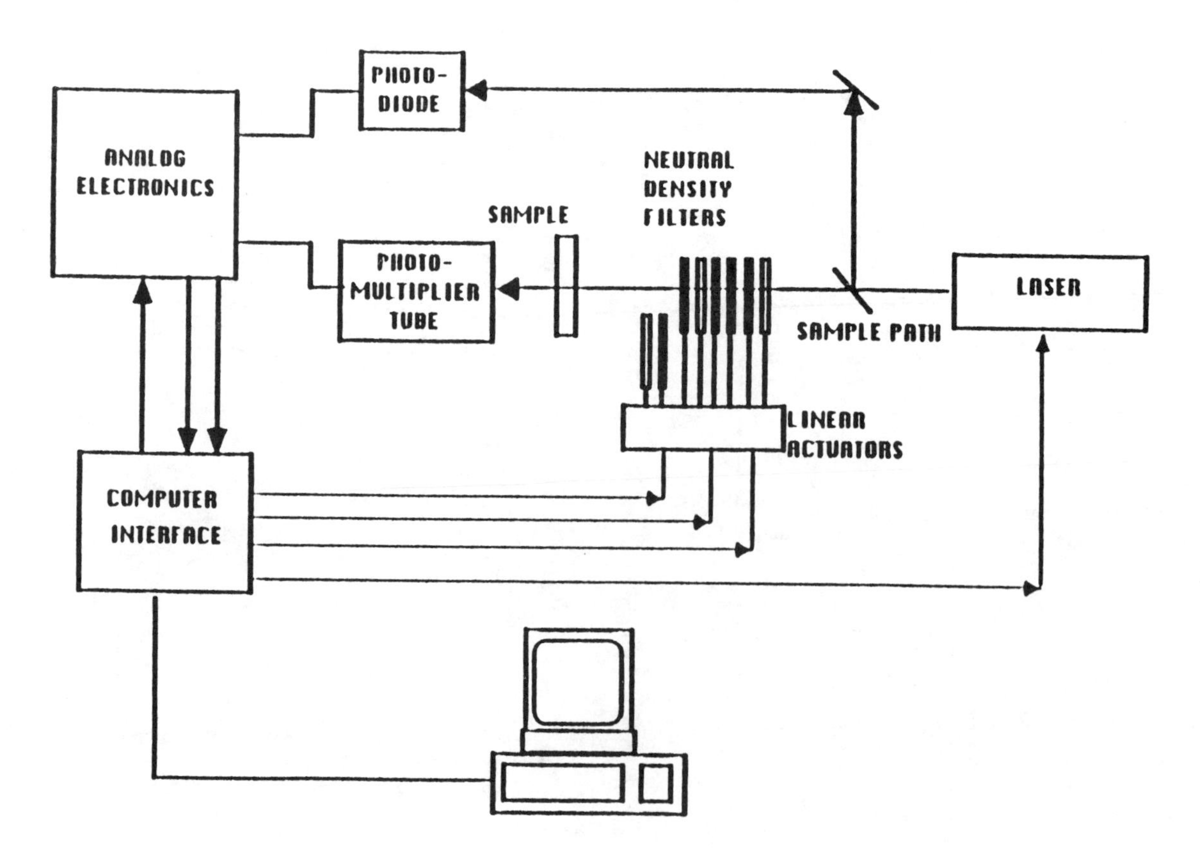

Figure 4. CW Laser Densitometer System Overview.

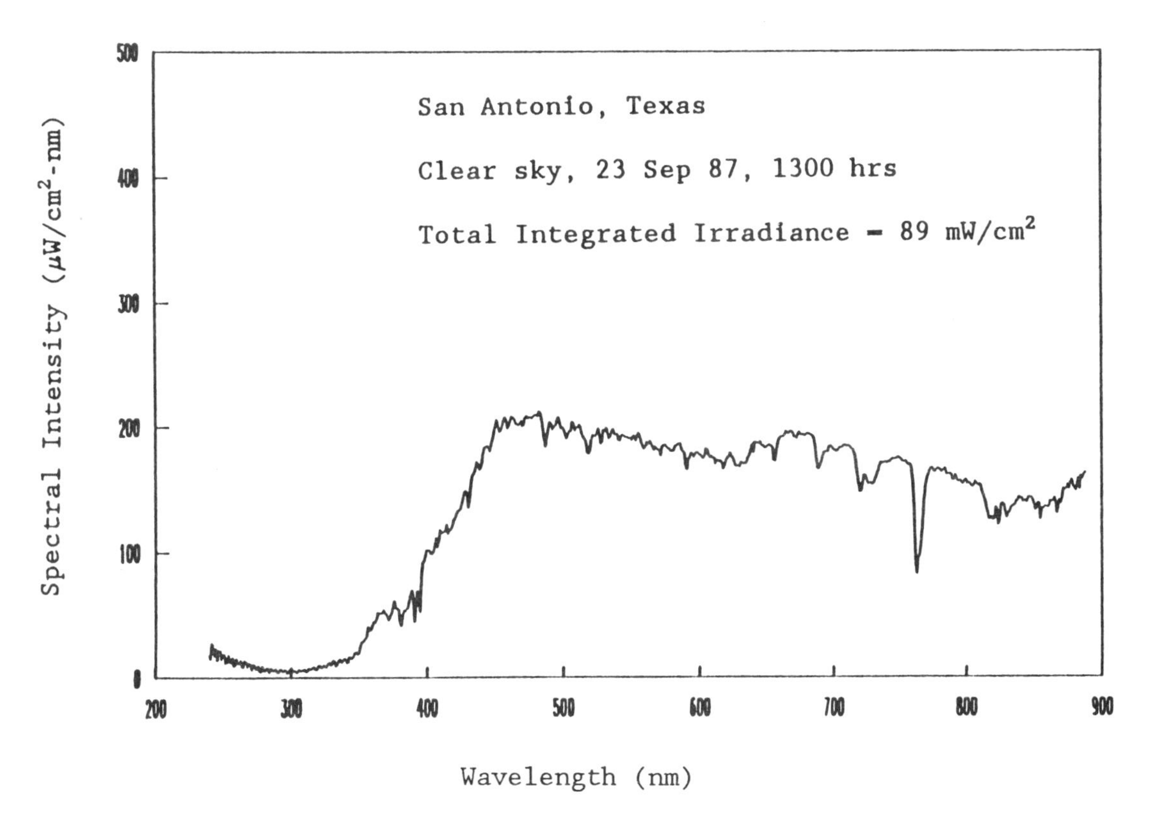

Figure 5. San Antonio Solar Spectral Irradiance.

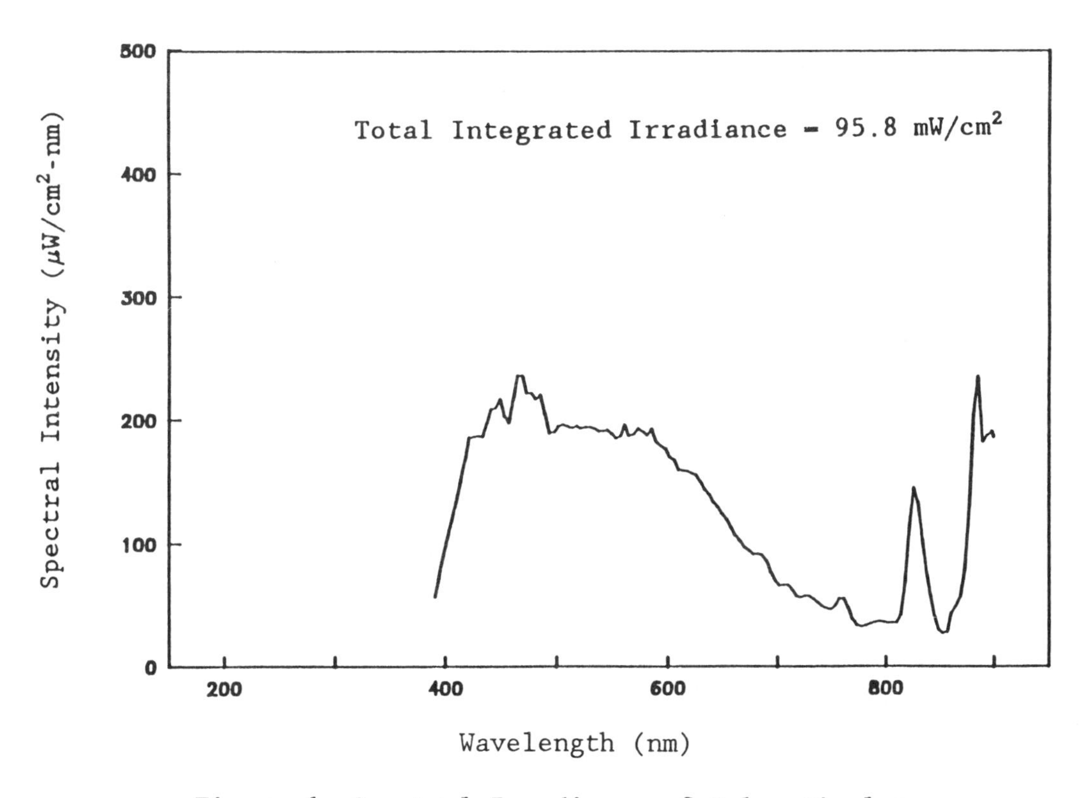

Figure 6. Spectral Irradiance of Solar Simulator.

HOLOGRAPHIC FILTER FOR COHERENT RADIATIONS

Dexiu Shi
Xiaozheng Xing
Myron L. Wolbarsht

Department of Biomedical Engineering, Duke University
Durham, North Carolina 27706

ABSTRACT

It is proposed to construct holographic devices which act as optical elements to modify the path of high spatially coherent radiation but not affect incoherent radiation. The device be composed of a laser-read holographic element in association with a band-pass spatial filter. The holographic filters will be wavelength independent to allow possible action as agile filters for laser safety protection while maintaining sufficient transmission for visual performance.

1. INTRODUCTION

With the growing applications in industry, medicine, science & technology, and even in the battlefield where the laser weapons can be used to blind enemy visual systems, the laser protection, particularly laser eye protection, is drawing more and more attention. A number of approaches have been explored to provide eye protection against laser radiation in the most common laser wavelengths. The laser eye protectors are available in the form of spectacles, dual lens goggles, coverall goggles or even a face mask. The objective of laser eye protection is to filter out the specific laser wavelength while transmitting as much visible light as possible so that the wearer can perform his or her task.[1] As laser radiation is produced in near monochromatic bands, it is possible to use band-cut filters to provide effective protection for a specific laser source without obstructing visual performance. There remain two major problems in practical laser protection. First, the commercially available laser eye protectors are designed for just few specific lasers and normally are not compatible with each other. With the use of newer laser wavelengths, special materials and tests will be required. Second, the premise of knowing which lasers are being used and which filter needs to be worn makes the band-cut filter almost useless in battlefields. Although it is easy to overcome this problem by employing a wide-spectrum-cutoff filter which has to be effective at more than one scene or for the tuneable lasers at any wavelengths, such an effective filter would have no visible transmission.

It is the purpose of this paper to discribe the construction of a versatile protective device in connection with laser eye safety. The principle upon which the protection is based depends on the degree of coherence required of a beam in reconstructing a hologram. A laser-read hologram will be expected not to affect incoherent radiation which can be transmitted providing the visual performance. Such holographic filter consists of two functional parts. The first is a lensless Fourier transform hologram of a laser source. The hologram produces a zero-order spectrum and a pair of first-order spectra when this hologram is reconstructed by the coherent laser source, while non-coherent light is minimally affected and uniformly attenuated. The second is an intensity band-pass spatial filter by which both the first-order and the zero-order spectrum of laser light in Fourier transform plane can be eliminated, and the non-coherent light in the environment. Therefore, this combined optical system will be wavelength independent to provide eyesafe protection with normal visual performance.

2. BASIC PRINCIPLES

It is apparent that what is recorded on the photographic plate is the interference pattern due to the two waves. The requirement placed on light sources to be used for hologram formation is those special properties which are known collectively as "coherence", and separately as spatial and temporal coherence.[2,3]

2.1 Spatial and temporal coherence

Spatial coherence is the degree to which a beam of light appears to have originated from a single point in space. Coherent illumination is obtained whenever the light appears to originate as a point source. The most popular example of a source of such light is the laser. Spatial coherence can be generally expressed as

$$\mu_s = \frac{\int_0^\infty V(r_1, t)\, V^*(r_2, t)\, dt}{\left[\int_0^\infty V(r_1, t)\, V^*(r_1, t)\, dt \int_0^\infty V(r_2, t)\, V^*(r_2, t)\, dt\right]^{1/2}} \quad (1)$$

where $V(r_1, t+\tau)$ and $V(r_2,t)$ represent the complex electric fields at points r_1 and r_2, and τ is the transit time difference for light traveling from r_1 and r_2 to an observation point. The quantity μ_s is called the complex degree of spatial coherence and its modules is the degree of spatial coherence. It is well known that a laser oscillating in a single mode has perfect spatial coherence with $|\mu_s| = 1$, and a laser oscillating in many transverse modes has only partial spatial coherence, i.e., $|\mu_s| < 1$.

Spatial coherence is inversely proportional to the apparent diameter of the source. The degree of spatial coherence $|\mu_s|$ of an uniform non-laser source depends on its radial extent:

$$|\mu_s| = \left| \frac{J_1(2\pi r_0\theta/\bar{\lambda})}{\pi r_0\theta/\bar{\lambda}} \right| \tag{2}$$

where r_0 is the source radius, $\bar{\lambda}$ is the mean wavelength, and θ is the angular extent of the source wavefront. When the source is an ideal point so that in Eq.(1) $r_0 = 0$, the degree of spatial coherence has the maximum value of one.

The temporal coherence of a light source is ultimately determined by the spectral purity of its radiation. Knowing the spectral width, it is possible to calculate the distance over which the photons remain substantially in phase, and this distance is called the coherence length. For white light it is only a few hundred nanometers; but for a laser for which the wavelength spread is very small, it may be from several centimeters up to kilometers. It is the coherence length of a light source that determines the available depth of the subject space in making a hologram, and a limit on the offset angle between object and reference beams.

2.2 Coherence limitations for hologram reconstruction

There are two main types of hologram classified in reconstruction coherence limitation: those which must be reconstructed using laser light or at least light exhibiting partial coherence; and those which can be viewed using white light. The image-plane holograms are made with objects wholly in the plane of the emulsion and can use white light that is not spatial coherent. The reason is that the distance over which the light must remain in phase is effectively zero[2]. However. the real interest in our approach is to make a laser-read hologram in which the image can be viewed only by coherent laser (or some kind of quasi-monochromatic) light.

If an original wavefront recorded in a hologram is to be reconstructed with minimum aberration, the direction and the radius of curvature of the reconstructing beam must duplicate those of the reference beam used to form the hologram. The resolution of the image produced by the wave reconstructed under this condition is limited only by the size of hologram and the coherence properties of the reconstructing light source. There is a simplified formula to describe the limitation of spatial coherence degree for hologram reconstruction.[3]

$$\Delta s = (z_i/z_p) r_0 \tag{3}$$

where Δs is the image resolution, z_i and z_p are the distance from image and reconstruction source to hologram, respectively, and r_0 is the source diameter which determines the degree of spatial coherence of the illuminating source. Eq.(3) implies that the further the image is from the hologram, the higher the degree of spatial coherence required from the illuminating source. On other hand if the ratio of z_i/z_p is fixed, for example $z_i/z_p = 1$, only the light of high degree coherence can be the reconstruction source.

2.3 Lensless Fourier transform hologram.

For the more general geometries for recording holograms and reconstructing images, the locations of two images in wavefront-reconstruction systems which use spherical reference wave and spherical reconstruction wave are given by the forms[4]

$$z_i = \left(\frac{1}{z_p} \pm \frac{\lambda_2}{\lambda_1 z_r} \mp \frac{\lambda_2}{\lambda_1 z_o}\right)^{-1} \tag{4}$$

$$r_i = \mp \frac{\lambda_2 z_i}{\lambda_1 z_o} r_o \pm \frac{\lambda_2 z_i}{\lambda_1 z_r} r_r + \frac{z_i}{z_p} r_p \tag{5}$$

where λ_1 and λ_2 are recording and reconstructing wavelength, respectively; z_i, z_o, z_r and z_p are the distance from the images, object, reference source and reconstruction source to hologram film, respectively. Similarly, here r_i, r_o, r_r and r_p are polar coordinates of the images, object, reference source and reconstruction source, respectively.

The relation among the directions of the object point source S_o, reference source S_r at recording, the images S_i and the source of illumination S_p at reconstructing is given by[5]

$$\theta_i = \left(\frac{\lambda_2}{\lambda_1}\right)(\pm\theta_o \mp \theta_r) + \theta_p \tag{6}$$

Figure 1 gives the geometrical descriptions of those parameters. In Eq.(4)-(6) the upper set of signs applies for one wave and the lower set for the second wave. When z_i is positive, the image is virtual; when z_i is negative, the image is real. If the different wavelengths are used, only the magnification of the image is changed. The magnification depends on the ratio of the wavelength used in viewing to the wavelength used in making the hologram, in addition to the source distances.

Differentiating the both sides of Eq.(5), we have

$$dr_i = \frac{z_i}{z_p} dr_p \tag{7}$$

where dr_i and dr_p represent the resolution of the reconstructing image and source diameter. It should be noticed that Eq.(7) is identical with Eq.(3).

For a particular interest case, when z_o and z_r are the same distance from holographic plate resulting in $z_i = z_p$, we have what is called a lensless Fourier-transform hologram. With $z_p = \infty$, the twin images generated by the hologram itself both lie at infinite distance regardless of the wavelength ratio λ_2/λ_1. If the reference source lies on the z axis, then the images are symmetric about the z axis. Alternatively, this hologram can be thought of as a sinusoidal amplitude grating, and when illuminated with monochromatic, spatially coherent light, the hologram produces a zero-order spectrum and a pair of first-order spectra.

3. CONFIGURATION OF A COHERENT RADIATION FILTER

Based on the above principle, it is possible to design a holographic image filter in such a way as to allow reconstruction of the image or, otherwise, give optical modification by and to the coherent portion of the light only. This holographic filter provides coherent laser protection with two functional processes. The first process is to modify the path of high spatially coherent radiation and leave the non-coherent light uneffected which can be accomplished by a lensless Fourier-transform hologram of a laser source. When this hologram is reconstructed by the coherent laser source, the hologram produces a central image and two auxiliary images at infinite distance from the hologram which correspond to a zero-order spectrum and a pair of first-order spectra in Fourier transform plane. Non-coherent light is minimally affected and uniformly attenuated.

The experimental configuration for making a laser transmission hologram of a laser source is shown in Fig.2(a). The incident laser beam which consists of spatially and temporally coherent light is divided into two beams which should be at approximately the same intensity. Both beams pass through a lens-prism combination optical system. One is taken as object beam (OB) and another as reference beam (RB). The optical path is carefully arranged with $z_r = z_o$. As evidenced by Eq.(4), when this laser source hologram is illuminated by a reconstructing coherence laser source which can be regarded as a plane wave with $z_p = \infty$, then the hologram produces a zero-order spectrum and a pair of first-order spectra as shown in Fig.2(b). This hologram is equivalent to a sinusoidal amplitude grating for collimated light. The diffracting angle is proportional to the ration of λ_1 and λ_2.

The Second process is to eliminate the coherent radiation from the reconstructed image and retain the non-coherent light for visual performance. Spatial filtering is an operation to remove (or preferentially pass) certain desired spatial frequencies. Various kinds of filter are encountered in spatial frequency filter techniques, and we are interested in the band-pass spatial filters. A schematic diagram of a typical spatial filter suitable for use in an optically aided viewing system is shown in Fig.3. A band-pass spatial filter is placed behind the laser-read hologram. This optical filter allows the certain band range of spatial frequencies to pass through while the higher and lower frequencies or the zero-order spectrum are stopped, and its the simplest form may be a transparent ring on a opaque substance as shown in Fig.3. The cutoff frequencies of the filter are, of course, determined by the aperture parameters of the iris. Therefore, an appropriate band-pass spatial filter can be used to remove the coherence radiation by which both the first-order and the zero-order spectrum of laser light in Fourier transform plane will be obstructed and the non-coherent light will be allowed pass through partially. In conclusion, this combined optical system will be wavelength independent to provide eyesafe protection with normal visual performance. This holographic element would thus act as a passive agile filter.

4. DISCUSSION

We have presented a holographic filter design that can reject the laser radiation to provide effective laser eye protection. The main advantages of this filter are its wavelength independence of laser protection and potential of visual performance. However, the geometrical parameters of experimental configuration need to be carefully optimized and the hologram processing techniques also need to be improved. Moreover, this filter works primarily for the normal incident laser light. Further study is proposed to improve the above system into a more general type suitable for a range of angles of incidence laser by the aided collimation of viewing or the convolution of the hologram construction procedure.

5. REFERENCES

1. D. Sliney and H. Le Bodo, "Laser Eye Protectors," *J. Laser Applications*, Vol.2, No. 3 & 4, pp. 9-13, Summer/Fall 1990.
2. G. Saxby, Practical Holography, Chapter 4, Prentice Hall, New York and London, 1988.
3. R. J. Collier, C. B. burckhardt and L. H. Lin, *Optical Holography*, Chapter 2 & 7, Academic Press, New York, 1984.
4. J. W. Goodman, *Introduction to Fourier Optics*, Mcgram-Hill Book Inc., San Francisco and New York, 1968.
5. M. Francon, *Holography*, Chapter 3, Academic Press, New York and London, 1974.

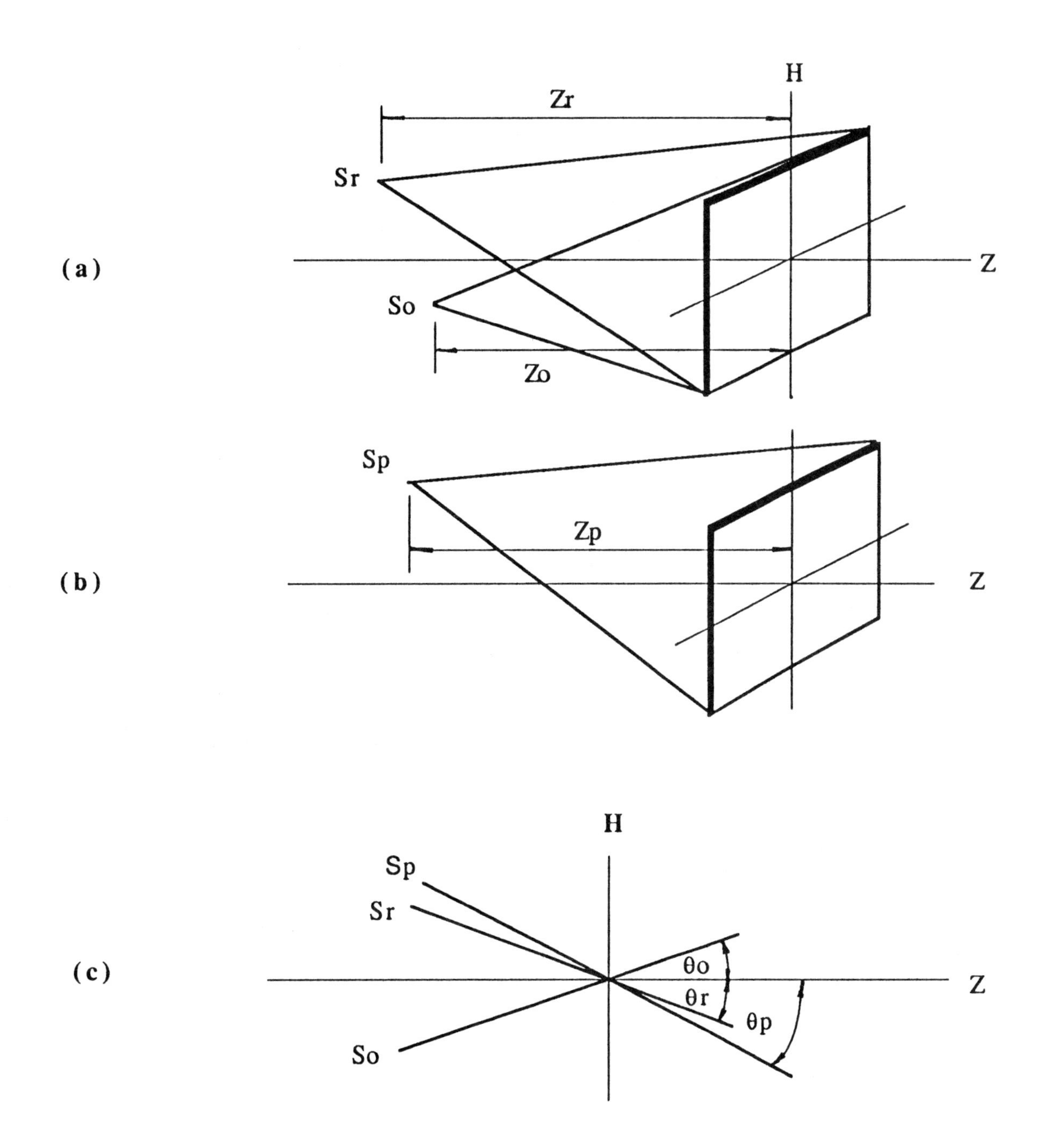

Figure 1. General geometries for recording holograms and reconstructing images with spherical waves: (a) recording, (b) reconstructing, (c) diagram of the directions.

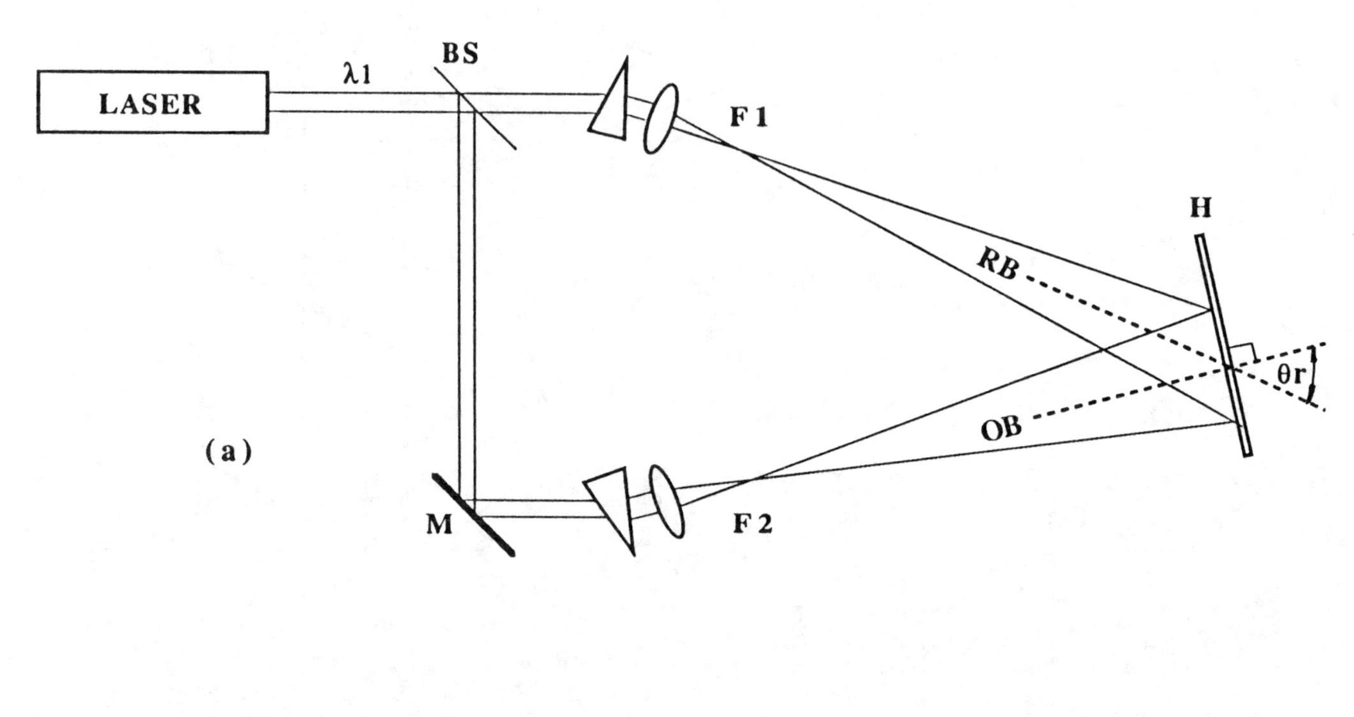

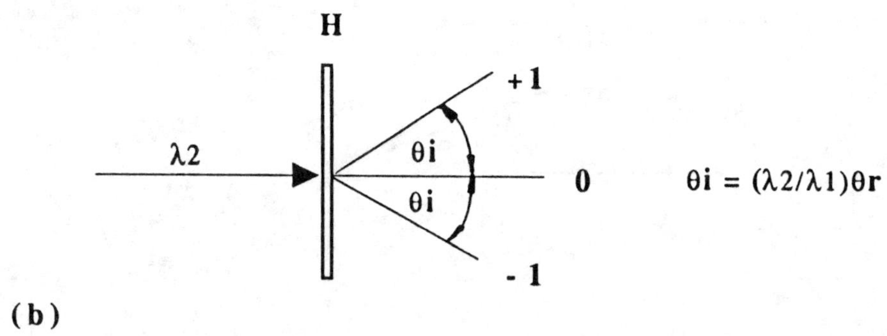

Figure 2. Laser transmission hologram. (a) Shows the table geometry; the focal points F1 and F2 should be equidistant from the hologram H. (b) Shows the hologram used as a sinusoidal grating.

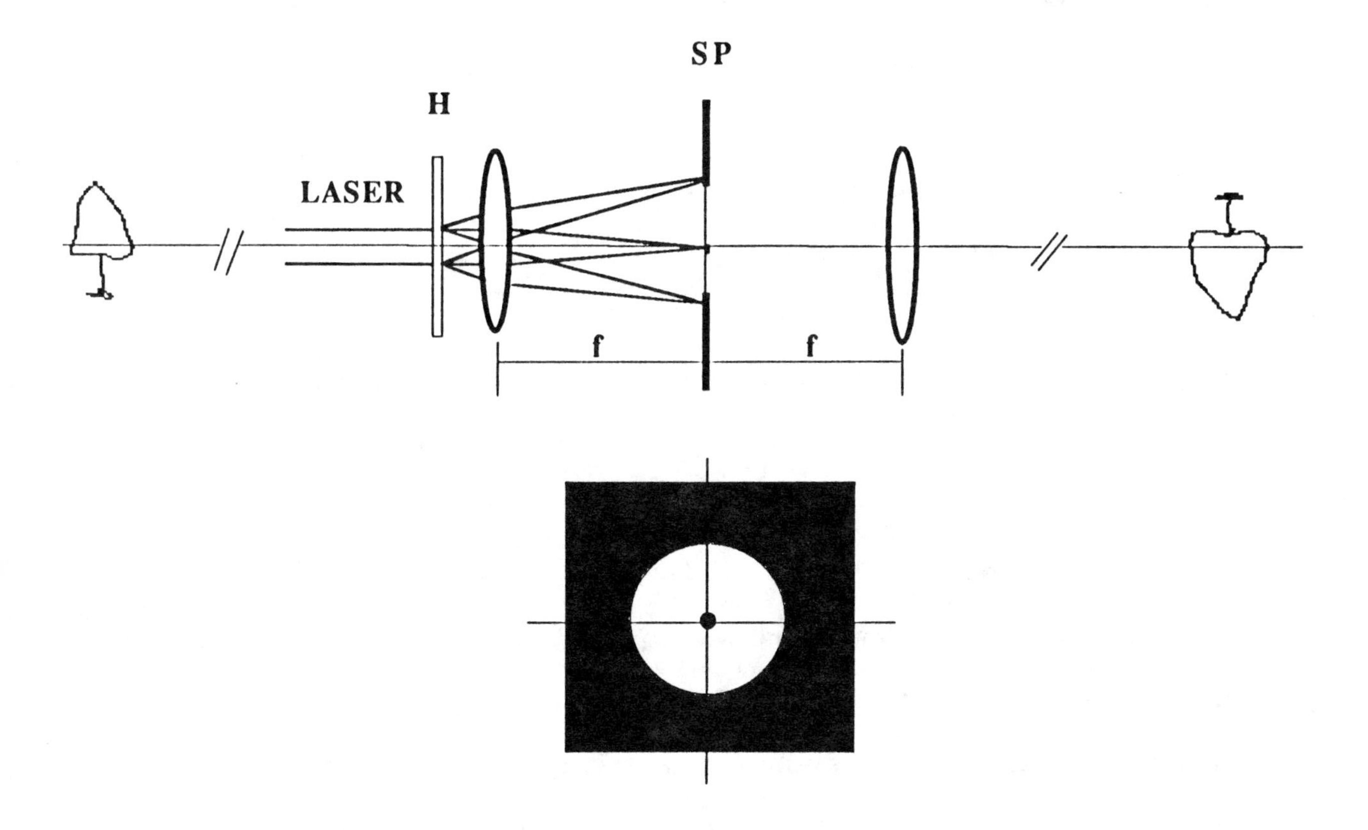

Figure 3. Schematic diagram of the holographic filter for coherent radioation. A band-pass spatial filter is positioned behind the laser transmission hologram. The laser light can be eliminated, while the incoherent background light will be transmitted in an attenuated form.

PASSIVE Q-SWITCHING OF EYESAFE Er:GLASS LASERS

B.I.Denker, G.V.Maksimova, V.V.Osiko, A.M.Prokhorov,
S.E.Sverchkov, Yu.E.Sverchkov, Z.Gy.Horvath.[+]

General Physics Institute, Vavilov str., 38, box 117333, Moscow, USSR. Phone (095)135-02-16, Telex 411074 LIMEN SU, Fax (095)135-02-70

[+]Central Research Institute for Physics, H-1525, box 49, Budapest, Hungary.

The search for simple and effective methods of erbium glass lasers (lasing wavelength 1.54 microns) Q-switching is a task of present interest. It is interesting because this wavelength radiation is relatively eyesafe and may be used in medicine, lidars, and fiberoptics. Nowadays, existing active methods of Q-switching are rather inconvenient for practical use (rotating prism) or complex and not too efficient (electrooptic cell). As for passive shutters based on color centers in crystals, or organic dyes, nowadays they are only under development, and Q-switching is obtained in them only as a small effect.

In this report, two new simple methods of passive Q-switching of erbium glass lasers are suggested, and the main properties of obtained giant pulses are investigated.

The peculiarity of erbium laser glasses is a pure 3-level laser scheme. It means that their luminescence spectrum in the 1.5 micron region is practically the same as their absorption spectrum. Thus, erbium glass laser light is absorbed in erbium-doped glass of the same or similar composition, and the cross section of absorption is close to that of emission. It is also clear, that after exciting 50% of Er ions the absorption will be saturated. In the case of comparable cross sections of stimulated emission and absorption focusing of the laser light is needed to form a giant pulse. That is because the energy needed to open the shutter should be small in comparison with the energy stored in the laser rod. So we used a laser resonator with light focusing into the shutter, as shown in Figure 1. Experiments with this scheme were made using a laser rod 100 mm long and 6.3 mm in diameter made of a new Nd-Yb-Er glass LGE-N developed in our institute. The rod was placed in a diffusively reflecting ceramic cavity and was cooled by heavy water. Output mirror reflected 62% of light. The shutter was a thin (several hundred microns) glass plate of the same composition as laser rod, but doped with erbium in concentration of 7×10^{20} ions per cubic centimeter. The thickness of the beam waist was regulated by adjusting the distance between the lens and the curved mirror. This adjusting should make the laser resonator stable enough not to be much influenced by thermooptical distortions of the rod, but should provide sharp enough focusing of laser light into the shutter.

The best results were obtained with a shutter plate with initial transmission of 82%, situated in the beam waist at Brewster angle. An interesting property of the laser was its spontaneous operation at TEM_{00} mode even if a low optical quality laser rod was used. The mode size was some 1 mm with a pulse energy of 6 mJ. We consider this phenomena is caused by mode selection by the soft aperture formed when the shutter opens. In the case when optical losses for TEM_{00} mode were especially made greater, for example at high pulse repetition rate, a multimode lasing was obtained. In this case, laser radiation filled all the aperture of the rod. The pulse energy was some 130 mJ. Pumping energy was some 145 J. At such pumping, free run output energy was 350 mJ. An optimized mini-laser based on a small (3 mm thick and 50 mm long) Cr-Yb-Er glass rod emitted 6 mJ single mode and 10 mJ multimode pulse at pump energy of 20 J.

One of the demands for the shutter glass is that Er ions exited stated lifetime in it should be small in comparison with the time needed to restore inversion of the laser glass. Typically this time is 300 - 500 micro seconds. If it is not the case, and the shutter remains opened long after the giant pulse and the pulse can be followed by free running. In the glass we designed for passive shutters, erbium luminescence was quenched down to 35 microseconds. This was found to be quite enough to suppress free running.

Typical giant pulse structure is presented in Figure 2. One can see it consisting of 1 - 2 (sometimes 3) separate spikes, each having halfwidth of some 70 ns. The delay between spikes is several hundred nanoseconds. We explain such repetitive spiking as follows: if the first spike inhomogeneously filled the rod aperture, there will remain highly populated zones in it. And as the shutter remains opened for some 30 micro seconds, an additional spike can be emitted.

In most practical cases such pulse structure is not desirable. To avoid repetitive spiking one should make the first and the only Q-switch pulse fill all the laser aperture and depopulate the rod homogeneously. It was found out that the easiest way to do so is to insert some kind of depolarizer into the resonator, or simple to operate at appropriate pulse repetition rate, when thermal depolarization occurs in the rod.

In the next part of the report we suggest another simple method of erbium glass laser passive Q-switching. It utilizes the property of semiconductors to reflect better when exposed to intensive light.

In our experiments we used a polished germanium plate, which served as a non-transparent mirror and an erbium glass laser rod described earlier. The best results were obtained with 62% reflectivity mirror. In this case, free run threshold was 175 J and at 177 J pumping several free run spikes were observed followed by a giant pulse. It had some 150 mJ energy and 100 ns duration. Alas, but the pulse damages the semiconductor surface, so this simple Q-switching method provides only a single use.

Summary

Two new effective methods of passive Q-switching for erbium glass lasers are suggested: Q-switching by an erbium containing glass with severely quenched luminescence and by germanium mirror. Both methods were experimentally tested and peculiarities of lasing described.

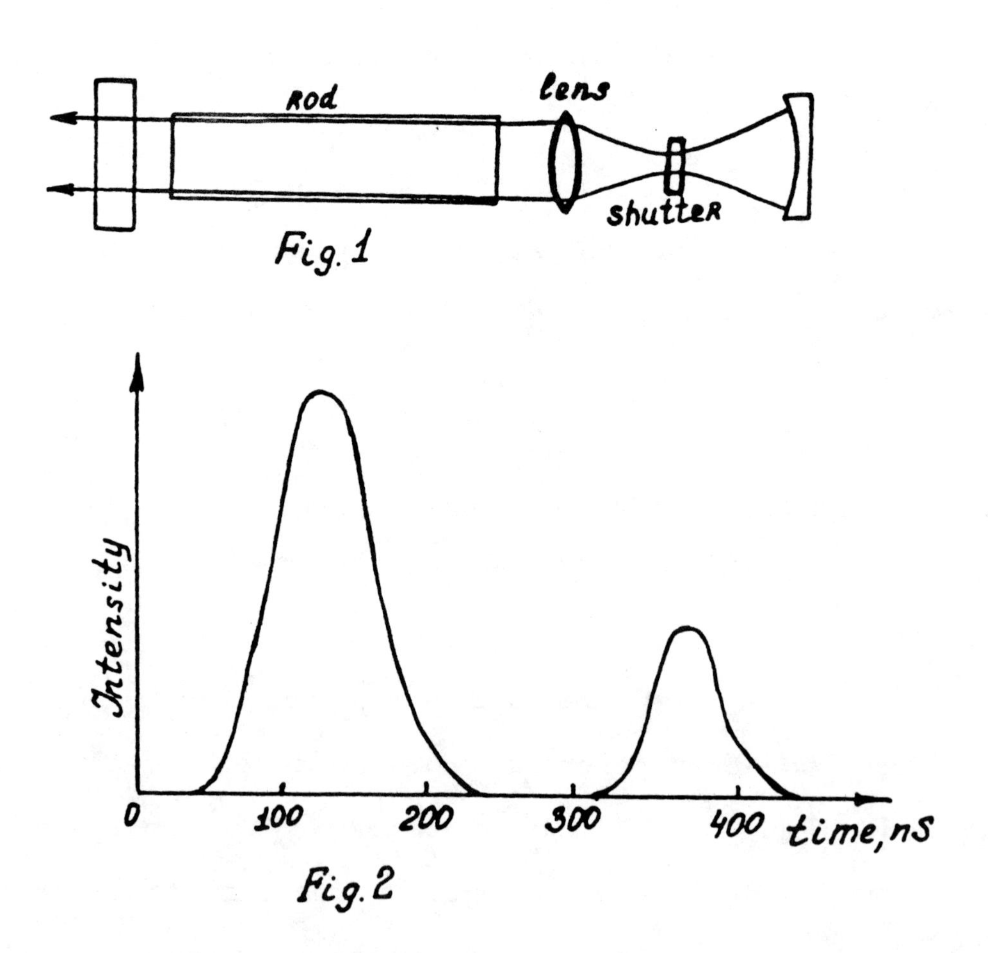

Fig. 1

Fig. 2

Energy storage efficiency and population dynamics in
flashlamp pumped sensitized erbium glass laser

Matjaž Lukač
ISKRA - Electrooptics, Stegne 7, Ljubljana 61210, Yugoslavia, and
Jozef Stefan Institute, Jamova 39, Ljubljana 61000, Yugoslavia

ABSTRACT

Erbium population dynamics in ytterbium sensitized phosphate glass is studied by measuring transient changes in laser probe transmission during flashlamp pumping. The influence of energy loss channels such as excitation cummulation and nonlinear fluorescence quenching on the flashlamp pumping efficiency is observed to be relatively small. The decrease in the pumping efficiency at high input energies can be attributed mostly to the shift in the flashlamp output radiation spectrum.

1. INTRODUCTION

One of the limiting factors in the operation of Q-switched Yb sensitized erbium glass lasers is the depletion of the erbium $^4I_{13/2}$ metastable upper laser level as a result of the excitation cummulation transfer[1] through the process $Yb(^2F_{5/2} - {}^2F_{7/2}) \rightarrow Er(^4I_{13/2} - {}^4F_{9/2})$, and, particularly at higher erbium concentrations, nonlinear quenching of the luminescence[2] from the Er $^4I_{13/2}$ level. Investigations into energy storage regularities have shown that the accumulation efficiency depends strongly on the composition of the glass.[1] To study the influence of energy loss channels and sensitizers on the accumulation efficiency in flashlamp pumped (Nd,Yb,Er): phosphate glass Kigre QE 7 we have carried out measurements of the erbium energy storage efficiency and transient population dynamics under various pumping conditions.

The energy level scheme relevant to the Er^{3+} pumping is shown in Fig. 1.[3] The laser action of Er^{3+} at 1.54 μm is produced by the transition $^4I_{13/2}$ - $^4I_{15/2}$. The fluorescence of higher states is practically fully quenched by the process of non-radiative multiphonon relaxation which has a rate of 10^5-10^7 s^{-1}.[1] Besides direct pumping of erbium absorption bands, the excitation of erbium is in QE-7 glass carried out through the sensitizer ions $Yb^{3+}$and Nd^{3+}. Our rough measurements[4] indicate that QE-7 glass contains approximately 1. 10 $^{19}cm^{-3}$ Er ions, 1. 10 $^{19}cm^{-3}$ Nd ions, and 1. 10^{21} cm^{-3} Yb ions. The Yb^{3+} ion absorbs into the wide $^2F_{5/2}$ band (0.9-1.1μm) from where a transfer occurs to the $^4I_{11/2}$ band of Er^{3+}. The Yb $^2F_{5/2}$ band can also be pumped by the transfer from the $^4F_{3/2}$ band of Nd^{3+} which has been populated by decay from the absorption bands of Nd^{3+} in the visible.[3]

2. EXPERIMENTAL

Cylindrical 3mm by 50 mm erbium glass laser rod was pumped by a 3mm bore, 40 mm arc length xenon flashlamp with a cerium doped quartz envelope. The flashlamp pulse forming network (PFN) provided critically damped pulses with adjustable pulse durations t_p within a 0.1 -2.1 msec range. The glass cylindrical pumping cavity with the outside dimensions of 15 mm by 43 mm, and the wall thickness of 1mm, was on the outer surface coated with chemically deposited silver. Changes in the populations of erbium $^4I_{13/2}$ and $^4I_{15/2}$ levels were monitored by measuring changes in transmission of a weak probe argon laser light. The transmitted argon laser light was isolated from the strong flashlamp light with a single-grating monochromator. During flashlamp pumping, the transmission of the 488nm argon line undergoes a transient increase, and the transmission of the 473nm argon line undergoes a transient decrease. From the known Er^{3+} spectra we conclude that the 488 nm and 473 nm lines excite the $^4I_{15/2}$ -$^4F_{7/2}$, and $^4I_{13/2}$ -$^2K_{15/2}$ erbium transitions. respectively. It is assumed that only these two transitions are excited, and that the contribution of other erbium lines to the laser probe absorption is relatively small. The 488 nm line thus probes the $^4I_{15/2}$ ground erbium laser level, and the 473 nm line probes the $^4I_{13/2}$ excited laser level. Our observations show that the peak transmission at 488nm (which corresponds to the minimum of the erbium ground state population) and the minimum transmission at 473nm (corresponding to the maximum of the erbium $^4I_{13/2}$ level population) occur within the experimental error of 50 μs at the same time. This is in agreement with the published multiphonon relaxation rate between the erbium $^4I_{11/2}$ and $^4I_{15/2}$ levels of 10^5 sec^{-1}.[5]

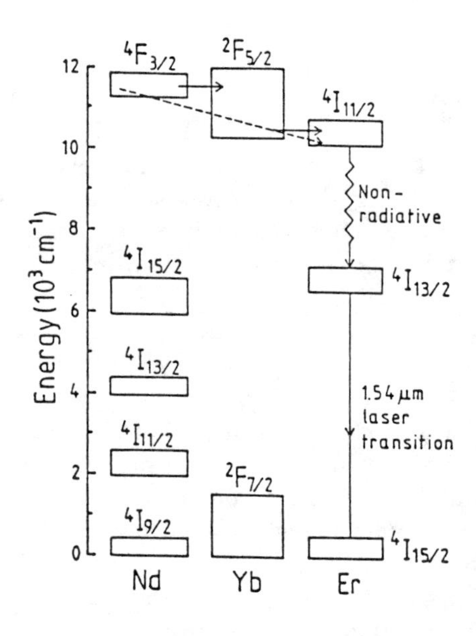

Fig. 1: Energy level and transfer diagram for lower levels of Nd^{3+}, Yb^{3+}, and Er^{3+} ions in a glass host.

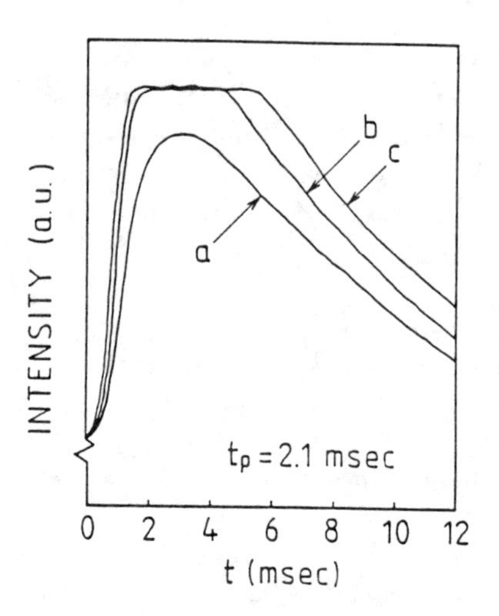

Fig. 2: Transient changes in the transmitted intensity of the 488 nm probe light during flashlamp pumping with pump pulse of duration t_p=2.1 msec for a) E_{in} = 30 J, b) E_{in} = 70 J, and c) E_{in} = 100 J.

Figure 2 shows typical transient changes in the transmitted intensity of the 488 nm argon laser line following the triggering of a flashlamp pulse of the duration t_p= 2.1 msec for several levels of pumping, E_{in}. The input energy E_{in} is determined by the relation $E_{in} = C.U^2/2$, where C is the discharge capacitance, and U is the initial voltage on the capacitor. When the population dynamics is monitored by the 488 nm line, the population inversion $n=(n_2-n_1)/n_o$ (where n_1 and n_2 represent the $^4I_{15/2}$, and $^4I_{13/2}$ erbium ion levels , respectively, and $n_0 = n_1 + n_2$) can be related to the transmitted intensity I of the probe light by the relation

$$n(t) = 2 \frac{\ln\left(\frac{I_{min}}{I(t)}\right)}{\ln\left(\frac{I_{min}}{I_{max}}\right)} - 1 \quad . \qquad (1)$$

Here, I_{min} is the observed transmitted intensity through the unexcited material (i.e. when n =-1), and I_{max} represents the transmitted intensity for the case when all Er^{3+} ions are in the upper laser level (i.e., when n=1). Figure 2 shows that in the case of t_p =2.1 msec, the transmission gets saturated for pump energies above appr. 70 J. In our work we assume that this saturated transmission, I_s, corresponds to the case when all Er ions are excited, i.e. we assume that $I_{max} = I_s$.

To distinguish contributions of different pumping bands to the erbium energy storage, the flashlamp output radiation was spectrally filtered with plane parallel glass filters inserted in the pump cavity between the flash lamp and the laser rod. As an example, Figure 3 shows the transient transmission changes for two pumping conditions: a) the pump light filtered with a cut-off RG 840 filter which absorbed shorter wavelengths below the transition wavelength of λ_f =840 nm, and b) the pump light filtered with a KG-3 filter transmitting most of the visible light but absorbing spectrum above approximately 700nm. It can be seen that pumping through the RG 840 cut-off filter is approximately 5 times more efficient than through the KG-3 filter. This is in agreement with our previous observation[4,6] that the ytterbium pump band between 0.9 - 1.1 μm is in QE-7 glass responsible for as much as 85 % of the total erbium inverted population energy storage. Figure 3 also shows that with the pump light above 840 nm, the peak population inversion is reached at approximately 0.7 msec after the flashlamp pulse has decayed. This indicates a relatively slow Yb - Er energy transfer. Our previous measurements have also shown that pumping in the 645-750 nm region is responsible for as much as 10% of the heat deposition in the laser rod while it's contribution to the pumping efficiency is negligible. Since there is a strong Nd^{3+} pump band in this region (See Fig. 1) this indicates that neodymium sensitization is inneficient.[7]

3. THEORY

A number of simplifying assumptions will be made in order to describe Er^{3+} upper laser level pumping. Since our measurements show that most of Er^{3+} ions are excited via Yb^{3+} sensitizer ions, we neglect direct erbium ion, and indirect neodymium sensitized pumping. We assume that the multiphonon relaxation between Er^{3+} states $^4I_{11/2}$ and $^4I_{13/2}$ is faster than the sensitization process. In this case all of the Er ions are in either excited laser level 2 or ground level 1. Since $^4I_{11/2}$ Er^{3+} level is assumed to be empty, we ignore the $Er(^4I_{11/2}$ - $^4I_{15/2}) \rightarrow Yb(^2F_{7/2}$ - $^2F_{5/2})$ back transfer process.[1] The Yb $^4F_{5/2}$ and $^4F_{7/2}$ levels will be represented by the population densities N_2 and N_1, respectively. The evolution of the normalized population inversion densities $n=(n_2-n_1)/n_o$, and $N=(N_2-N_1)/N_o$, where $N_o=N_2+N_1$, is described by the following rate equations:

$$\frac{dn}{dt} = W_s\,(1+N)\,(1-n) - \frac{(1+n)}{\tau_{Er}} \tag{2a}$$

$$\frac{dN}{dt} = -\,r\,W_s\,(1+N)\,(1-n) - \frac{(1+N)}{\tau_{Yb}} + D\,(1-N)\,\left(\frac{t}{0.4\,t_p}\right)\exp\left(\frac{t}{0.4\,t_p}\right), \tag{2b}$$

Here, τ_{Er} and τ_{Yb} are respectively the life-times of the Er $^4I_{13/2}$ and the Yb $^4F_{5/2}$ excited states. The process of sensitization is described by an average constant sensitization rate $W_s(sec^{-1})$. The last term in Eq. (2(b)) describes flashlamp pumping of ytterbium which is assumed to vary in time proportionally to the flashlamp discharge current;[8] t_p is the flashlamp current pulse duration as measured at the 10% pulse height. It is assumed that $N_2 << N_1$. The proportionality constant D measures the strength of the pumping, and depends among other factors on the flash lamp input energy and output radiation spectrum. The parameter r represents the ratio of Er^{3+} and Yb^{3+} ions, $r= n_o/N_o$. Note that in Eqs. (2) we ignored excitation cummulation as well as nonlinear quenching. This simplification will be justified by the relatively good agreement between experiment and the model.

In our model a constant rate W_s is assigned to the Yb-Er energy transfer process. In reality, a wide dispersion of collection rates by Er^{3+} sites from Yb^{3+} donor surroundings exists. At large interionic distances energy is transferred through the multipole electric interactions. On the other hand, our investigations[9] into ion pair interactions in other rare earth doped systems indicate that the coupling between the closest ions is larger than expected from the simple dipole-dipole interaction, and probably is due to an exchange interaction. As a result of the dispersion in collection rates, the Er^{3+} ions with higher rates are excited initially and then do not participate in the transfer process due to lack of energy migration in the erbium sub-system. This effect which can lead to a fast decrease in the sensitization rate,[1] is ignored in our model.

When pumping processes are short compared to τ_{Er}, the change in n can be described by:

$$\frac{dn}{dt} = W_p\,(1-n)\quad , \tag{3}$$

where $W_p(sec^{-1}) \equiv W_s\,(1+N(t)$, is the time dependant pumping rate. The maximum inversion density, n_{max}, that can be obtained with a single pump pulse can then be expressed as:[4]

$$n_{max} = 1 - 2\exp(-A) \quad , \tag{4}$$

where A represents the pump integral defined as the integral of the pumping rate over the pump pulse duration:

$$A = \int W_p \, dt . \tag{5}$$

Note that when a laser is operating in a free oscillation mode the population inversion increases during pumping only until it reaches the laser threshold value n_t. In situations when the total pump integral is larger than the threshold pump integral A_t defined by:

$$A_t = -\ln\left(\frac{1 - n_t}{2}\right) \quad , \tag{6}$$

the remaining difference $(A - A_t)$ is used up for the laser oscillation. It can be shown[4] that when the pump integral A is a function only of the flashlamp input energy but not, for example, of the population inversion or laser photon density, a following relation exists between the total pump integral A and the output energy E_{out} of the laser amplifier :

$$E_{out}(E_{in}) = K\,(A(E_{in}) - A_t) \quad , \tag{7}$$

where

$$K = \tfrac{1}{2} V\, h\, \nu\, n_o \frac{\gamma_1}{\gamma} (1 - n_t) \quad . \tag{8}$$

Here, V is the laser material volume, $h\nu$ is the laser transition photon energy, γ is the total fractional photon loss in a single roundtrip passage inside the resonator, and γ_1 represents the fraction of photons emitted through the output laser mirror. Both, the proportionality parameter K, and the threshold pump integral A_t can be calculated from the basic laser material and resonator properties. The dependence of the total pump integral A on the input energy can thus be obtained by measuring the laser output vs. input energy characteristics.

4. RESULTS

The evolution of n as obtained from the measured transmitted intensity variations are for several flashlamp input energies, and for the pulse duration of $t_p = 150\mu s$ shown in Fig.4. The solid lines represent the theoretical fit on n(t) obtained by numerical integration of Eqs. (2). The following values are used in the fit: $\tau_{Yb} = 3$ msec,[2] and $r = 1.\ 10^{-2}$. The best agreement with the data for all pumping strengths is obtained with $\tau_{Er} = 7.5 \pm 0.1$ msec, and $W_s = 2.3\ (1 \pm 0.1)\ 10^5\ sec^{-1}$. Note that the population decay at different erbium inversions is adequately described by a single erbium $^4I_{13/2}$ level decay time indicating that non- linear quenching is not important at erbium concentrations of $1.\ 10^{19} cm^{-3}$. A slightly lower decay time than observed in other erbium glasses with no Nd sensitizer might indicate that a back transfer to the $^4I_{15/2}$ of Nd occurs.[3]

Our data is not exact enough to confirm any existence of higher initial sensitization rates due to the nearest Er-Yb neighbor interactions. However, Fig. 4 shows that the sensitization process can be described quite adequately by the average constant sensitization rate W_s. Approximately 10^{-2} Yb^{3+} ions need to be excited in order to achieve full erbium population inversion. The rate of the erbium inversion change, W_p, is therefore on the order of $10^3\ sec^{-1}$.

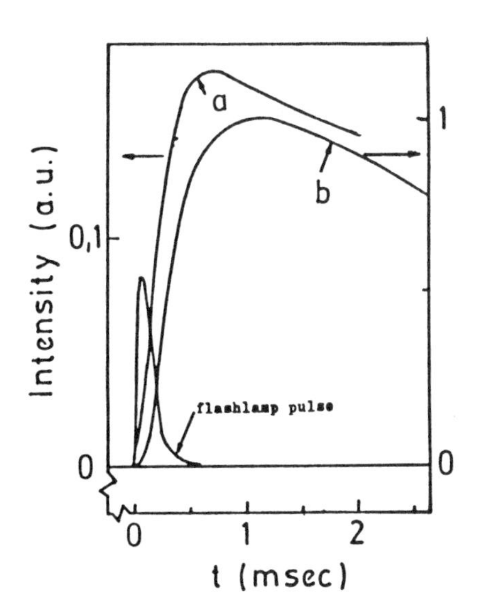

Fig.3: Transient transmission changes for pumping through a) KG-3, and b) RG-840 filter; t_p = 0.15 msec, and E_{in} = 57 J.

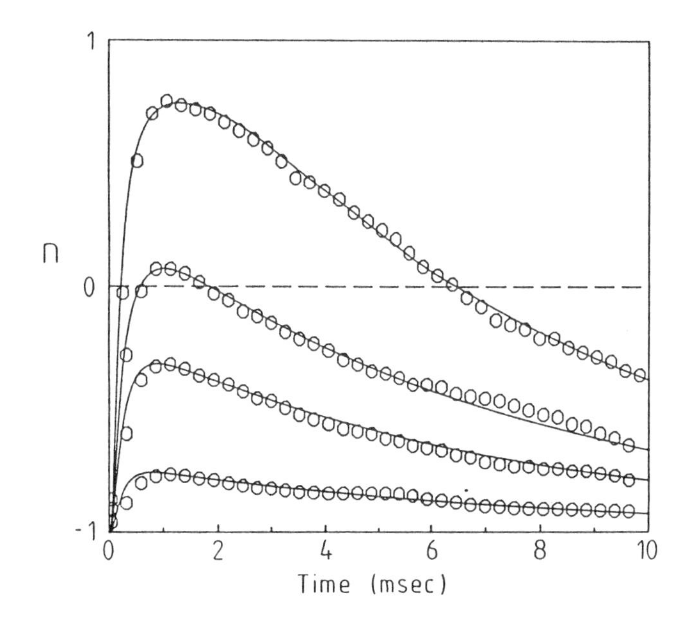

Fig. 4: Measured (circles) and calculated (full line) temporal evolution of the erbium inversion during flashlamp pumping for several flashlamp input energies; t_p= 0.15 msec.

When the strength of pumping, D, is a linear function of E_{in}, our model assumes that the maximum inversion is related to the input energy by

$$n_{max} = 1 - 2\exp(-a E_{in}) \quad , \tag{9}$$

where a is a proportionality constant. In our experiment, the maximum inversion is determined by measuring the maximum transmission during the pumping transient. By measuring maximum inversion for different input energies it is possible to check the validity of Eq. (9) which takes into account only the depletion of the ground state. A deviation of the experimentally measured relation from that assumed by Eq. (9) would indicate either that D is a nonlinear function of E_{in}, or that at higher erbium inversions, loss processes occur, such as excitation accumulation or nonlinear quenching. In order to distinguish between these two effects, we have carried out output energy measurements of the erbium laser operating in a free-oscillation mode.[4,6] Figure 5 shows the measured output energy curves for different flashlamp pulse durations. No lasing bellow E_{in} = 60 J could be obtained for the shortest pulse duration in our experiment of t_p = 0.1 msec. As can be seen, the pumping efficency drops at short pulse durations, and at high input energies. Since erbium inversion remains constant during lasing this decrease in the pumping efficiency cannot be explained by the excitation accumulation or nonlinear quenching. We explain this effect by the dependence of the output radiation spectrum on the flashlamp power density. Namely, a total pulsed lamp's spectral output distribution is determined by the power density J defined by $J=E_{in}/(t_pS)$ where S is the internal surface integral of the lamp's discharge region.[8] Both, line radiation due to discrete transitions between bound xenon energy states, and continuum of radiation due to the recombination of free-electrons and xenon ions, are present in a pulsed xenon flashlamp spectral output. Line spectra in the near infrared are more dominant at lower power densities while at high power densities the continuum radiation in the visible dominates. This is demonstrated in Fig. 6 which shows the ratio of the flashlamp output radiation spectral distributions $R(\lambda)$ integrated over the pulse duration for the pulse durations of t_p = 2.1 msec, and t_p = 0.1 msec. Since erbium ions are pumped mostly through Yb ions which absorb in the near infrared region, low flashlamp power densities with output spectrum shifted to longer wavelengths are more efficient.

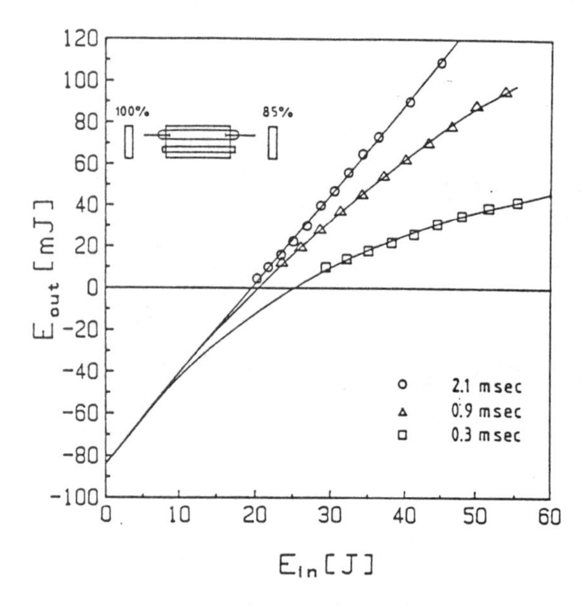

Fig. 5: Output energy data of QE-7 glass laser for different flashlamp pulse durations t_p. The optical resonator consisted of two flat mirrors with a a separation of 14.5 cm, and the output mirror reflectivity of 85 %.

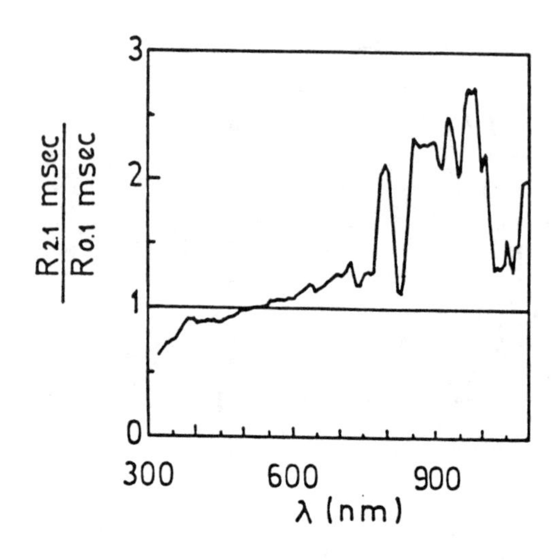

Fig. 6: Ratio of the flashlamp output radiation spectral distributions R(λ) for the pulse durations of t_p = 2.1 msec and t_p = 0.1 msec.

As can be seen from Fig. 6, the functional dependence of the output energy, and therefore of the pumping area A, can be approximated by a linear function only at low power densities. In particular, for the input energy range 0 - 50 J, the linear relationship exists only for the long pulse duration of t_p = 2.1 msec. It is only for these pumping conditions that the maximum inversion is expected to increase with the input energy according to Eq. (9).

The measured maximum inversions, as obtained from transmission transients, are for the pulse duration of t_p=2.1 msec, and t_p=0.1 msec shown in Fig. 7. Also shown are the predicted maximum inversion curves as calculated from the measured output energy characteristics. It is evident that the measured values for t_p = 2.1 msec agree well with those predicted by Eq. (9). Within the experimental error, no appreciable deviation can be observed at high inversions, indicating that the effect of excitation cummulation is relatively small. On the other hand, the shift of the pump spectrum to lower wavelengths at high power densities results in a much less efficient pumping with t_p = 0.1 msec long pulses. For this pulse duration, the measured erbium maximum inversion is for input energies up to 60 J below the particular laser resonator threshold inversion of n_t= 0.4. This is in agreement with the observation that no lasing could be obtained with t_p = 0.1 msec long pulses.

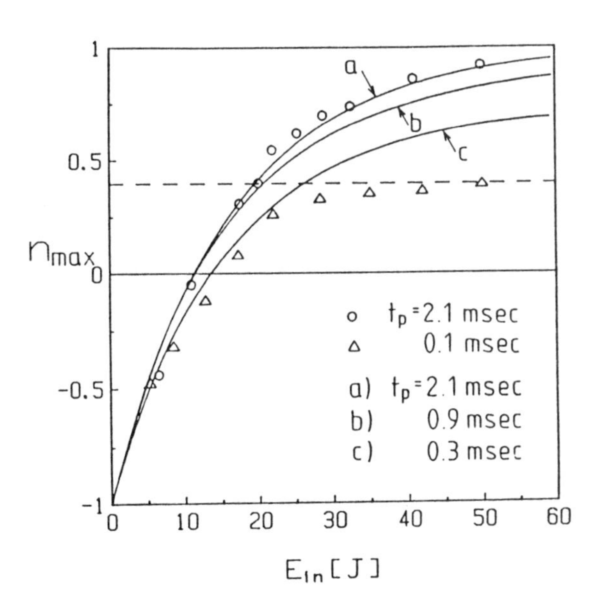

Fig. 7: Measured maximum erbium inversions as a function of the input energy for the flashlamp pulse durations of t_p = 2.1 msec (circles), and t_p = 0.1 msec (triangles). Full lines represent theoretical curves for pumping with flashlamp pulses of duration t_p = 2.1 msec (a), 0.9msec (b), and 0.3 msec (c). The dashed line represents the laser threshold inversion of n_t = 0.4.

5. CONCLUSION

The energy transfer dynamics from Yb^{3+} to Er^{3+} in a phosphate glass was studied by measuring transient changes in laser probe transmission during pulsed flashlamp pumping. Measurements show that a simple rate equation model where the sensitization process is described by a single constant rate of $W_s = 2.3\ 10^5\ sec^{-1}$ adequately describes the pumping process. No evidence of the nonlinear fluorescence quenching is observed.

Our analysis also shows that the decrease in the pumping efficiency at high input energies can be attributed mostly to the shift of the flashlamp output spectrum to lower wavelenths at high flashlamp power densities while the contribution of the excitation cummulation and nonlinear fluorescence quenching to the efficiency decrease is relatively small.

REFERENCES

1. V.P. Gapontsev, S.M. Matitsin, A.A. Isinev, and V.B. Kravchenko, "Erbium glass lasers and their applications," Optics and Laser Technology, pp. 189-196, August, 1982.

2. T.T. Basiev, E.V. Kharikov, V.I. Zhekov, T.M. Murina, V.V. Osiko, A.M. Prokhorov, B.P. Starikov, M.I. Timoshechkin, and I.A. Shcherbakov, "Radiative and nonradiative transitions exhibited by Er^{3+} ions in mixed yttrium-erbium aluminum garnets," Sov. J. Quantum Electron., Vol. 6 (7), pp796-799, 1976.

3. J.G. Edwards, and J.N. Sandoe, "A theoretical study of the Nd:Yb:Er glass laser," J. Phys. D: Appl. Phys., Vol.7, pp 1078-1095, 1974.

4. M. Lukač, and M. Marinček, "Energy storage and heat deposition in flashlamp pumped sensitized erbium glass lasers," IEEE J. Q. Electr., Oct. 1990.

5. C.B. Layne, W.H. Lowdermilk, and M.J. Weber, "Multiphonon relaxation of rare-eart ions in oxide glasses, Phys. Rev.," Vol. 16, pp. 10-17, 1977.

6. M. Lukač, and M. Marinček, "Effect of sensitizers on flash lamp pumping efficiency and heat deposition in Er glasses," in High-power solid state lasers, ed. C.L.M. Ireland, SPIE Proceedings, March 90, The Hague.

7. Kigre has informed us that Nd ions have been removed from the QE-7 glass composition.

8. ILC Technology, "An overview of flashlamps and cw arc lamps", Technical Bulletin 3.

9. M. Lukač, and E.L. Hahn, "Spectroscopy of symmetry broken optical doublets in Pr^{3+}: LaF_3," Opt. Comm., Vol. 70, pp195-201, 1989.

SESSION 2

Eyesafe Laser Systems and Applications

Chairs
Milton A. Woodall II
Optic-Electronic Corporation

Penelope K. Galoff
U.S. Army Environmental Hygiene Agency

EYESAFE HIGH PULSE RATE LASER PROGRESS AT HUGHES

R. D. Stultz, D. E. Nieuwsma, E. Gregor

Electro-Optical and Data Systems Group, Hughes Aircraft Company
P. O. Box 902
El Segundo, CA 90245

ABSTRACT

We report on new eyesafe laser hardware developed at Hughes Aircraft Company. We also report on the development of a compact 20 Hz pulse repetition frequency (PRF) backward Raman configuration and the investigation of deuterium as an alternative to methane for high PRF eyesafe Raman lasers.

INTRODUCTION

Lasers operating at wavelengths and power levels that are safe to the eye are increasingly required for both commercial and military applications. The 1.54µm laser wavelength is of greatest interest for these eyesafe applications, as can be noted from the number of papers at this second annual conference on Eyesafe Laser Systems. At last year's conference[1], we provided a scientific introduction into the Stimulated Raman Scattering process, which Hughes Aircraft Company has chosen as the most cost effective and efficient means to generate high pulse rate laser emissions at the 1.54µm wavelength. We, also, provided a basis for our system trade-off considerations, which lead to the selection of the Raman shifted Nd:YAG technology, and an overview of several fielded laser rangefinder prototypes built by Hughes. This work culminated in the construction of an eyesafe High Reprate Raman Rangefinder Demonstrator (R3D) unit, described last year.

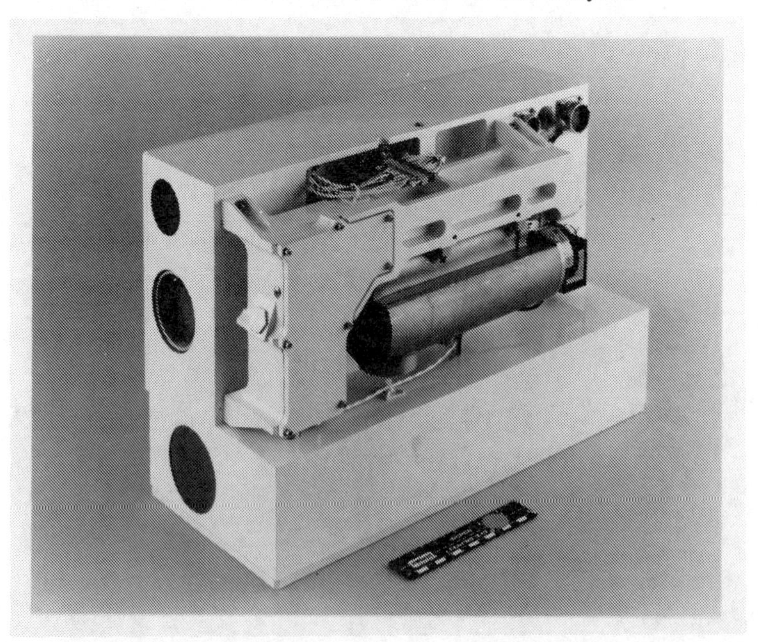

Figure 1. The High Reprate Raman Rangefinder Demonstration (R3D) system, fielded by Hughes in 1989, provides eyesafe rangefinding capability at PRFs up to 20 hertz, as demonstrated in several government field exercises.

The R3D, shown in Figure 1, was designed as a Raman shifted modification of a Nd:YAG air defense laser rangefinder previously built by Hughes and several of its licensees. The circulating gas Raman cell technology was added to this reliably proven Nd:YAG laser to produce an eyesafe rangefinder capable of operation at PRFs up to 20 hertz and 1.54μm laser energies of 40 milliJoules per pulse. This demonstration system was developed to verify the performance expectations of an air defense or airborne laser rangefinder. During the past year, the R3D has been successfully demonstrated through a series of tests conducted by the CECOM Center for Night Vision and Electro-Optics[2], at the Smoke and Obscurants Test at Eglin AFB in May 1990 (Smoke Week XII)[3], and at the Landing Aids demonstration conducted by NAEC in August 1990. These field tests have demonstrated eyesafe laser rangefinding against uncooperative targets under adverse weather and man made obscurants to distances of 10 kilometers. During these test programs, the R3D was operated for over 100 hours, consisting of nearly 2 million laser pulses, without degradation in performance or reliability.

In this current paper, we describe a fully militarized, high pulse rate, eyesafe laser rangefinder called the Electro-Optical Tracking Sensor (EOTS), which Hughes designed and built for an air defense gun system. This production laser system demonstrates the maturity and reliability of the Raman shifted Nd:YAG laser and InGaAs Avalanche Photodiode receiver components. We also discuss recent progress at Hughes in advanced technology for Raman shifting of laser emissions. Raman cells that take advantage of the phase conjugate properties of Backward Raman Scattering have been designed and tested which provide improved beam quality over conventional Raman shifting. In addition, tests have been conducted to demonstrate the lower refractive index distortion and greater thermal diffusivity that may be obtained by using Deuterium gas as the Raman shifting medium, instead of Methane. Nd:YAG Raman shifting in Deuterium provides 1.56μm laser light, and medium shows promise for much greater pulse rate operation, with a simpler Raman cell configuration.

HUGHES' ELECTRO-OPTICAL TRACKING SENSOR

In 1990, Hughes brought together the circulating gas Raman cell hardware developed over the past several years for high pulse rate, eyesafe lasing with an Eyesafe Receiver Hybrid to design and produce an eyesafe laser rangefinder for the Electro-Optical Tracking Sensor (EOTS) system. EOTS is a complete air defense fire control system, which integrates a Thermal Imaging Sensor and CCD TV with a high PRF eyesafe laser rangefinder on to a pointing and tracking gimbal equipped with a Stabilized Head Mirror Unit. Together with advanced image processing methods, EOTS is designed to provide a complete fire control solution for a vehicle mounted air defense weapons system.

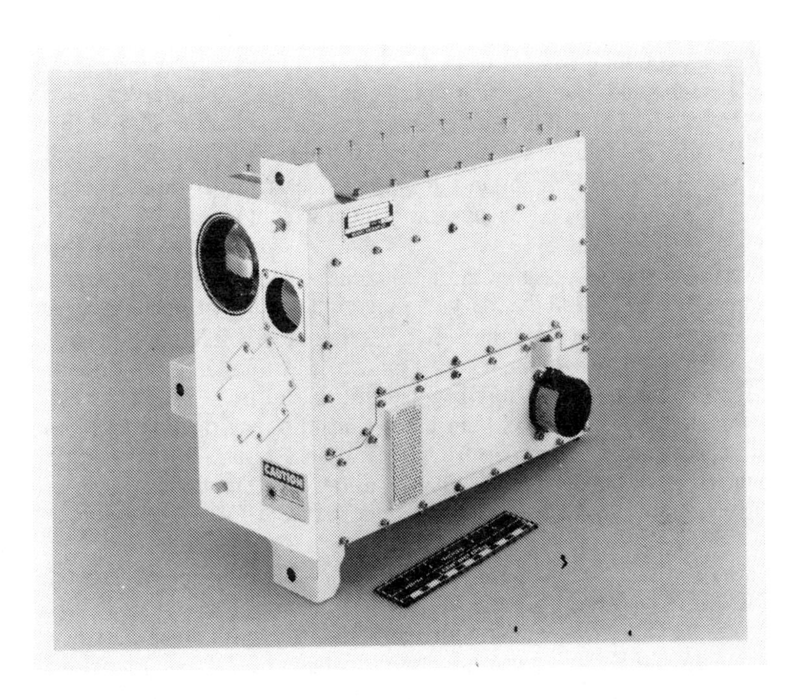

Figure 2. The Eyesafe laser rangefinder for the EOTS air defense system is a fully militarized device which combines field proven components into a single, small and reliable package.

TABLE I. EOTS Eyesafe Laser Rangefinder Performance Characteristics

Parameter	Value
Operating Wavelength	1.54μm
Output Laser Energy	35 milliJoules
Pulse Repetition Frequency	15 hertz
Duty Cycle (Run Time)	100% (Continuous)
Laser Pulse Width	14 nanoseconds (FWHM)
Beam Divergence	0.5 milliRadians
Output Beam Diameter	40 millimeters
Receiver Aperture Diameter	70 millimeters
Receiver Field of View	0.4 milliRadians
Detector NEP	< 2 nanowatts
Receiver Extinction Ratio	> 47 dB (500m, 85% target)
Maximum Range	9995 meters *
Minimum Range	50 to 750 meters (selectable)
Range Accuracy	± 5 meters
Multi-target Resolution	20 meters
Range Logic	First and Last Reply

* Adjustable to 19,995 meters or more.

The eyesafe laser rangefinder for the EOTS is based on the design for the High Reprate Raman Rangefinder Demonstration (R3D) unit. The EOTS Eyesafe Laser Rangefinder (EOTS ELRF), shown in Figure 2, is a military ruggedized single unit package, providing eyesafe laser transmitter, receiver and power supply all in one compact box. The performance characteristics for the EOTS ELRF are summarized in Table I.

The EOTS ELRF uses the "half resonator" Raman laser configuration[4,5], described in detail last year, as did the R3D system. In the "half resonator," a single optical flat acts as both the Nd:YAG output mirror and as the back mirror for the Raman resonator. There is no output mirror for the Raman resonator, since the focus in the Raman medium acts as a mirror. The use of the "half resonator" allows the design to use, without modification, many optical components previously proven on Nd:YAG laser production programs. For instance, the Nd:YAG laser rod pump cavity for this system is the same flashlamp pumped, pressurized gas cooling system used on over 3,000 production military lasers delivered by Hughes. Similarly, the electro-optical Q-switch and beam steering optics used are all based on previous designs. The Raman cell for the EOTS laser is a militarized version of the 10 inch, circulating gas Raman cell developed for the R3D system (see Figure 3). The transmitter optical path is folded four times; two folds in the Nd:YAG resonator, one through the Raman cell and a final fold out the beam expander; in order to maintain the system size at just over 1000 cubic inches (17,400 cm^3). By maintaining the dimensions at 13.6 by 11.5 by 6.8 inches, the system weight goal of 40 pounds (18 kilograms) is achieved.

Reliability of the EOTS receiver optics is assured through the use of a modification of the laser receiver telescope used for the M1 tank laser rangefinder. Over 8,500 of the M1 tank rangefinders have been delivered and the modifications for this eyesafe adaptation were demonstrated on an eyesafe rangefinder at Smoke Week XI, in February 1989[6]. The 1.54μm optical energy reflected off the target is collected by a 70mm receiver objective lens and directed through a narrow bandpass filter on to the APD detector diode. The high efficiency narrow bandpass filter transmits 90% of the energy within a ±25 nanometer bandwidth of the 1.543μm wavelength. With less than 0.1% transmittance outside this band, nearly all of the optical background noise that could degrade the receiver performance is blocked out. A visible reticule operating in the 0.65 to 0.70μm region is coupled to the receiver optical axis by a beam splitter near the diode and projected out the objective lens to provide an accurate line-of-sight reference that is displayed on either the CCD TV or Thermal Imager display.

Hughes has developed a hybrid receiver which utilizes an InGaAs Avalanche Photodiode (APD) for use in eyesafe rangefinder applications. The design was based on the M-1 tank receiver hybrid developed by Hughes. The new Eyesafe Receiver Hybrid incorporates the APD detector, along with a 35MHz bandwidth preamplifier, into the hybrid package. The gain verses bias curve for the InGaAs APDs are steeper than for the Silicon APDs commonly used with Nd:YAG rangefinders and hence require a more precise setting to obtain maximum gain with minimum noise. Thus, a more complex method than previously used for the Silicon APDs had to be developed for optimally setting the bias on the InGaAs APDs. Circuitry for automatically biasing the APD and automatically setting threshold levels to achieve maximum sensitivity has been included in the hybrid. Laboratory tests on the hybrid have shown less than 3dB degradation in sensitivity over the ambient operating temperature range of -32 to +50 degrees Celsius.

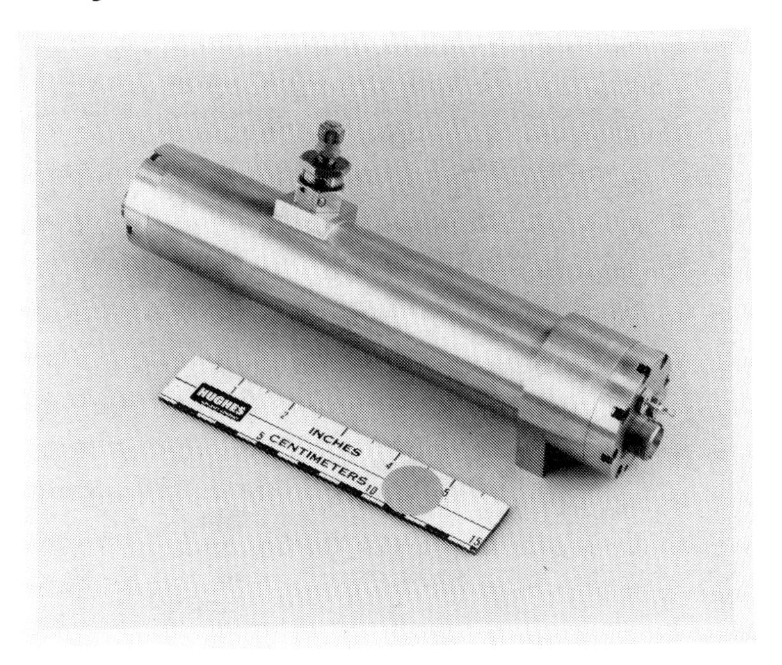

Figure 3. The compact circulating gas Raman gas cell, developed by Hughes for 1 to 30 hertz pulse rate operation, has been fully militarized for the EOTS application.

The receiver internally regulates the supply voltages to +5V,-10V, and -5V DC. The hybrid also requires +130V for supplying bias to the APD. It uses a TTL signal to initiate the auto-biasing and auto-thresholding sequences and a TTL signal approximately 750μseconds later to indicate the valid ranging time. All other logic functions are contained in the hybrid. The digital video output of the hybrid is complementary TTL. The Eyesafe Receiver Hybrid has been designed to run in both single shot and high PRF systems, up to a maximum PRF of 500 hertz.

Based on analysis of the performance of the rangefinder systems which have utilized the Eyesafe Receiver Hybrid, it is estimated to have better than 10 nanowatt sensitivity, for a 99% probability of detection, when used with a laser pulsewidth of 14 nanoseconds (FWHM). Analysis of the aforementioned systems indicate biasing of the APD at or near the designed current gain of 10 provides better than 10 kilometer ranging capability. Test data on the R3D and EOTS units have shown less than 1 false alarm per 1000 shots.

It was necessary to minimize the volume of the EOTS ELRF electronics in order to fit into the compact chassis with the optics. In order to accomplish this objective, the logic, control and interface functions are all performed by a single chip microprocessor. Running on a clock frequency of 11.0592 megahertz, it provides serial ports at 19.2 kilobaud for external communications. Input commands and output range data and system status are communicated by two 8-bit computer words over a RS422 bus.

The range counter was implemented in a configurable gate array. The gate array is configured to provide 5 meter range increments with a minimum range that can be set over the microprocessor port from 50 meters to 750 meters in fifty meter increments. The maximum range can be set to be 10 kilometers or 20 kilometers. The rangecounter provides first reply and last reply data on each ranging cycle.

The low voltage power supply for EOTS is a 2 inch by 3 inch hybrid producing up to 100 watts of output power. The supply runs at 500 kilohertz switching frequency and requires careful filtering external to the input to suppress the EMI inherent with this high frequency switching.

The PFN power supply design is an evolutionary step forward. The circuit type is still the flyback topography used in most of the laser systems produced by Hughes. This supply differs in the switching frequency and consequently the compactness achieved. The power MOSFET driven supply operates at about 100 kilohertz and is capable of producing 300 watts of output power. The power conversion efficiency is approximately 85% input to output.

In the EOTS system, the Eyesafe Laser Rangefinder is mounted vertically into a gimbaled tracker with a one way optical transmission of 80% and a pointing error less than 50 μradians (RMS). A computer model of the maximum range performance of this completed system has been generated. This model takes into account all system errors and assumes a target off-set from the EOTS line of sight equal to the 1/e angle of the laser beam divergence. The model has been used to predict the maximum range for a 99% per pulse probability of detection for several different targets under a variety of atmospheric conditions. The results of this model are best displayed in a graph of maximum range verses the atmospheric attenuation coefficient, as in Figure 4. In Figure 4, the results for two sample targets are presented. Both targets have a reflectivity of 10% at the 1.54μm wavelength. The atmospheric attenuation coefficient, calculated using PC-LnTRAN, for several weather conditions of interest is displayed in Table II. The atmospheres of Table II assume Mid-Latitude Summer conditions, per the Air Force model, but the graph of Figure 4 holds under any weather conditions. As is shown, the maximum range on a standard day with 23.5 kilometer visibility for the EOTS rangefinder is 13 km for a 4 square meter, 10% reflective target or 11.5 km for a 2 square meter target. If the visibility is 8 kilometers, the temperature 90°F and the relative humidity 80%, the maximum range is 9 km, for the 4 meter square target and over 8 km for the 2 square meter target.

The computer model used to generate Figure 4 was, also, applied to the conditions prevalent at the Hughes roof house laser test range in El Segundo, California. This range allows for controlled testing of laser rangefinders against a 75% diffuse reflective (at 1.54μm) target located at a distance of 590 meters. The computer model predicted the same receiver extinction ratio, to within 0.5dB, as was measured during roof house testing. This test provided an initial verification of the predicted performance of the Hughes Electro-Optical Tracking Sensor system. Further verification has been underway since last October during field testing of the first production configuration EOTS system as installed in an air defense vehicle. To date, the test program has verified the predicted results, including acquisition and tracking of high performance, fixed wing aircraft at distances well beyond 7 kilometers.

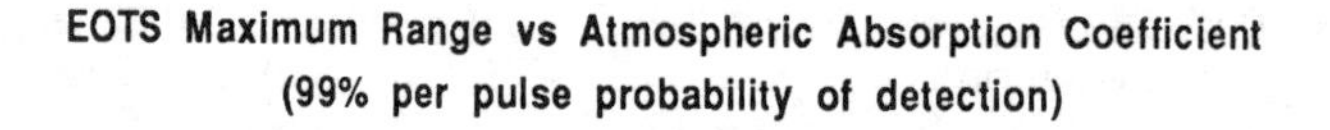

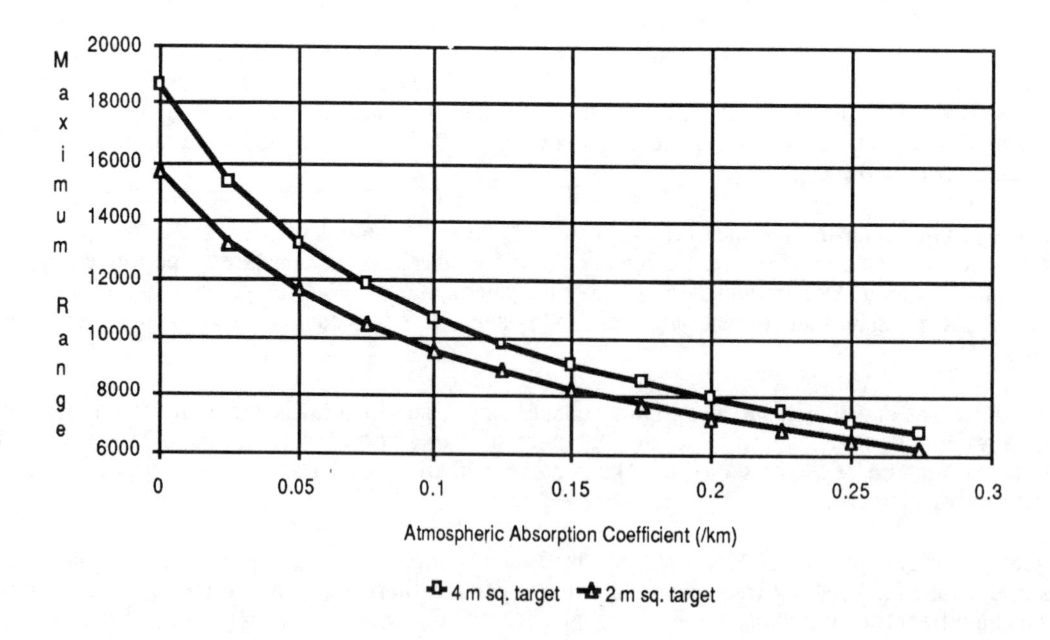

Figure 4. The maximum range in meters for a 99% probability of detection by the EOTS system is shown as a function of the Atmospheric Attenuation Coefficient at the 1.54μm wavelength. Two targets, both with a 10% reflectivity are displayed.

Table II. Atmospheric Attenuation Coefficients per Kilometer for Several Mid-Latitude Summer Conditions (calculated using PC-LnTRAN)

WEATHER CONDITION	VISIBILITY (kilometers)	TEMPERATURE	RELATIVE HUMIDITY	ATTENUATION COEFFICIENT
CLEAR	23.5	24°C	50%	0.055
CLEAR	8.0	32°C	80%	0.154
CLEAR	8.0	38°C	80%	0.167
CLEAR	8.0	32°C	100%	0.196
HAZY	5.0	24°C	50%	0.208
CLEAR	8.0	38°C	100%	0.212

STIMULATED RAMAN SCATTERING

Raman scattering is a two-photon, inelastic scattering of light[7]. Figure 5 is an energy level diagram for a molecule in the Raman medium. For our purposes here, we will be concentrating on vibrational Raman scattering in both methane and deuterium, so the initial and final states shown correspond to vibrational energy states of the molecule. Raman scattering is a process where a pump photon is absorbed and a Raman photon of lesser energy is simultaneously emitted leaving the molecule in an excited state. The frequency of the Raman photon is given by

$$\omega_2 = \omega_1 - \frac{\Delta E}{h} = \omega_1 - \omega_v \tag{1}$$

where ΔE is the energy difference between the initial and final states of the molecule. The Raman shift, ω_v, is a characteristic of the particular Raman medium used.

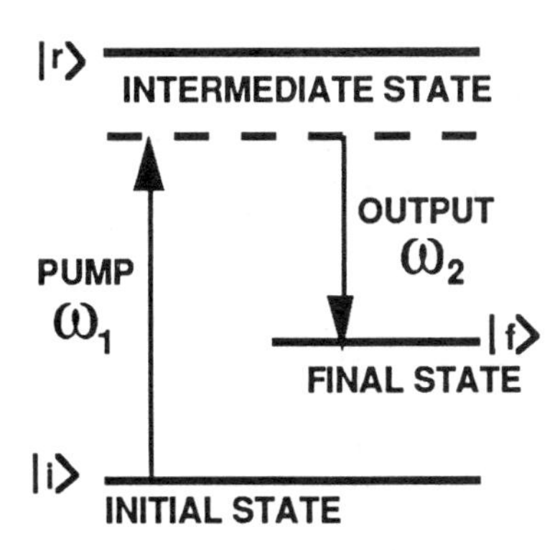

Figure 5. Raman scattering is a two photon process. In Stokes scattering (shown here) the incident photon (ω_1) is shifted to a lower frequency photon (ω_2).

At low pump intensities, only spontaneous Raman scattering occurs and therefore only an extremely small amount of Raman energy is produced. As the pump intensity is increased, the amount of Raman energy detected will at some point begin to increase nonlinearly due to stimulated Raman scattering (SRS). When pump depletion is negligible, the Raman intensity as a function of position z along the pump beam is given by

$$I_s(z) = I_s(0)\, e^{g I_p z} \tag{2}$$

where g is the Raman gain coefficient and I_p is the pump intensity. The gain coefficient depends on material parameters of the Raman medium[8]

$$g = \frac{\lambda_s^2 \lambda_p N}{c\, n_s^2\, h\, \pi\, \Delta\nu} \left(\frac{\partial \sigma}{\partial \Omega} \right) \tag{3}$$

where λ_s and λ_p are the Raman and pump wavelengths, respectively, N is the number density of molecules in the Raman gas, c is the speed of light in a vacuum, h is Plank's constant, n_s is the refractive index at the Raman wavelength, $\Delta\nu$ is the Half Width at Half Maximum (HWHM) Raman linewidth, and $\partial\sigma/\partial\Omega$ is the SRS cross section. In the case of a gas, $n_s \approx 1$. At a fixed temperature N is clearly pressure dependent, but so is $\Delta\nu$. In the pressure broadened regime, $\Delta\nu$ becomes proportional to N so that g is essentially independent of pressure. Table III has the cross section and gain coefficients[9] for methane and deuterium at the pressures indicated. The gain coefficient for deuterium is about 2.7 times smaller than that for methane.

Table III. Raman Scattering Cross Sections and SRS gain coefficients for Methane and Deuterium[9].

	$\partial\sigma/\partial\Omega$ (cm^2/Sr) X 10^{30}	Pressure (psig)	Raman gain coef. (cm/GW)
Methane	2.7	1000	1.2
Deuterium	0.79	1500	0.45

BACKWARD RAMAN

Vibrational stimulated Raman scattering (SRS) can occur in either the backward or forward direction relative to the pump beam propagation. We have experimentally found that backward Raman lasers have significantly improved Raman beam quality compared to forward Raman lasers. In some cases this can be a factor of 2 improvement. The reason for this may lie in the phase conjugate nature of the backward Raman beam[10]. It has also been reported that forward Raman beam quality can be degraded due to the coupling of the Stokes and anti-Stokes radiation[11]. This latter effect would not occur in backward Raman because of the large phase mismatch with anti-Stokes in that direction.

Since the backward beam is phase conjugate to the pump beam, the Raman beam quality and beam direction are very insensitive to misalignments or distortions of the Raman cell optics. This makes it easy to switch back and forth between the pump and Raman outputs when a 2 color laser system is desired.

Backward Raman lasers are very efficient. We have demonstrated greater than 50% conversion efficiency in methane with a 200 mJ pump. This corresponds to more than 72% quantum efficiency. The divergences and near-field diameters of the pump and output Raman beams were identical. Given the wavelength difference, this means that the Raman beam was 1.5 times closer to diffraction limit. We have also demonstrated 25% conversion efficiency in deuterium in a backward Raman configuration..

We have developed a 20 Hz PRF, 50 mJ @ 1.54μm backward Raman cell in a very compact form. The cell is identical in size to the EOTS Raman cell and can be easily fit into the EOTS transmitter. We achieved 40% conversion efficiency with this design along with a Raman beam divergence only 1.1 times the pump laser divergence.

DEUTERIUM INVESTIGATION

At present, methane is used exclusively to Raman-shift the Nd:YAG laser wavelength of 1.064μm to the eyesafe wavelength of 1.542μm. Methane has a Raman shift of 2915 cm^{-1}. Deuterium has a nearly identical Raman shift of 2986 cm^{-1} which converts the Nd:YAG wavelength to 1.56μm. This is slightly outside the very narrow official eyesafe band around 1.54μm; however, in actuality it could be just as eyesafe. Another possibility would be a 1.053μm Nd:YLF pump laser which would Raman-shift to 1.536μm in deuterium.

Methane Raman lasers have been demonstrated with several Joules of output energy[12], and at several hundred Hertz PRFs using flowing gas Raman cells. However, it is useful to investigate possible alternate eyesafe Raman gases with superior thermal properties. Deuterium is just such a gas and the ultimate goal of this investigation is to determine the feasibility of a 10-20 Hz PRF eyesafe Raman laser using uncirculated deuterium.

We will briefly mention here another advantage of using deuterium. Optical breakdown in methane has been known to create organic substances which ultimately lead to opaque deposits and damage on the Raman cell windows. This process will limit the lifetimes of methane cells unless steps are taken to avoid breakdown. Deuterium is a very simple molecule consisting of 2 heavy hydrogen atoms and therefore no organics are created in breakdown.

Because the thermal properties of deuterium are far better than those for methane, much higher PRFs in uncirculated deuterium should be possible. In order to gain some appreciation of this it is necessary to discuss how optical distortions are created in the gas following a laser pulse and how fast these distortions dissipate. We will show that useful figures of merit are [refractive index (n) - 1], and the thermal diffusivity (D).

The amount of energy deposited in the form of vibrational excitation is proportional to the Raman shift and to the amount of SRS energy produced. The excited molecules quickly convert their energy to thermal energy resulting in temperature fluctuations in the gas. This in turn creates density fluctuations which distort optical beams through the refractive index variations they cause. From the ideal gas law, the density (N) and temperature (T) are related by

$$NT = \text{constant} \tag{4}$$

for constant pressure. From (4) we obtain

$$\frac{\Delta N}{N} = -\frac{\Delta T}{T} \tag{5}$$

The refractive index is related to the density by the Lorentz-Lorenz relationship for an isotropic medium[13]

$$\frac{n^2 - 1}{n^2 + 2} = \frac{4\pi N}{3}\alpha \tag{6}$$

By differentiating both sides of (6) and using $n \approx 1$ for a gas we obtain

$$\Delta n \approx (n-1)\frac{\Delta N}{N} = -(n-1)\frac{\Delta T}{T} \tag{7}$$

This means that for a given ΔT, Δn is roughly proportional to (n-1). We see that the smaller the refractive index of the Raman gas, the smaller the optical distortions will be that result from temperature fluctuations.

The thermal relaxation time (τ) after which the gas returns to its equilibrium state will be approximately proportional to the inverse of its thermal diffusivity which is defined as

$$D = \frac{\alpha}{\rho\, c_p} \quad (8)$$

where α is the thermal conductivity, ρ is the density, and c_p the specific heat at constant pressure, i.e.

$$\tau \sim \frac{1}{D} \quad (9)$$

For laser interpulse periods much shorter than τ and without gas circulation, there is considerable optical distortion in the beam path which degrades Raman conversion efficiency and beam quality. The maximum PRF possible in an uncirculated gas should go as $1/\tau$ and therefore as the diffusivity D.

Table IV has the pertinent thermal and optical parameters for both methane and deuterium. These were determined for a pressure of 1000 psig methane and 1500 psig deuterium. (n-1) is about a factor of 2 lower, and the diffusivity a factor of 4 higher for deuterium. Thus, we expect the maximum PRF possible in deuterium to be about 4 times higher than that of methane for these pressures. 1000 psig is a typical operating pressure for methane eyesafe Raman lasers; however, the pressure chosen for deuterium was not optimum, and we expect that it can be reduced by at least a factor of 2 without reducing the Raman gain since it is in the pressure broadened regime. Reducing the pressure will make the diffusivity even higher, as can be seen from (8).

Table IV. Optical and Thermal Parameters of Methane and Deuterium.

	Pressure (psig)	(n-1) X 10^{30}	Thermal diffusivity (cm^2/sec) X 10^3
Methane	1000	30	3
Deuterium	1500	14	12

Although the prior analysis does yield some figures of merit to aid in choosing the best candidate for high PRF applications, it is rather crude and must be backed up with experiment. We have made some experimental measurements of the thermal relaxation time for both methane and deuterium. Figure 6 shows one possible experimental arrangement for making this measurement. The gas in the Raman cell is pumped with a Nd:YAG laser as shown. A helium-neon (HeNe) laser at 633nm is used to probe the distortions in the Raman gas. The HeNe beam passes twice through the cell and a portion of the reflected beam is directed into a fast photodiode detector using a beamsplitter. The detector has a small active area comparable to the size of the HeNe beam. Normally, the detector signal is high since virtually all of the reflected HeNe beam is incident on the detector. However, when the laser is fired, the distortions in the Raman gas cause the HeNe beam to fall off of the active area and the detector signal drops rapidly. After a while the distortions in the gas dissipate and the detector signal returns to its normally high state. The time it takes for the detector signal to recover is a measure of the relaxation time of the gas.

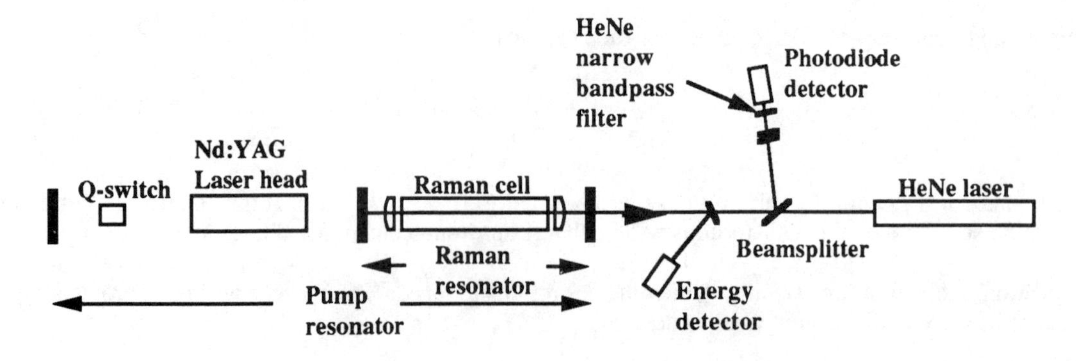

Figure 6. Thermal relaxation time experiment.

Some sample measurements are shown in figures 7a and 7b. With a 1.56μm output of 20 mJ, 1500 psig of deuterium had a relaxation time of about 150 ms. In a similar test, with an output of 20 mJ of 1.54μm, 1000 psig of methane had a relaxation time of about 1 second (6 or 7 times longer than for deuterium). We were able to comfortably operate the laser at 5 Hz PRF using uncirculated deuterium and an output of 20 mJ. However, a methane Raman laser cannot be operated much more than about 1 Hz without using some sort of gas circulation.

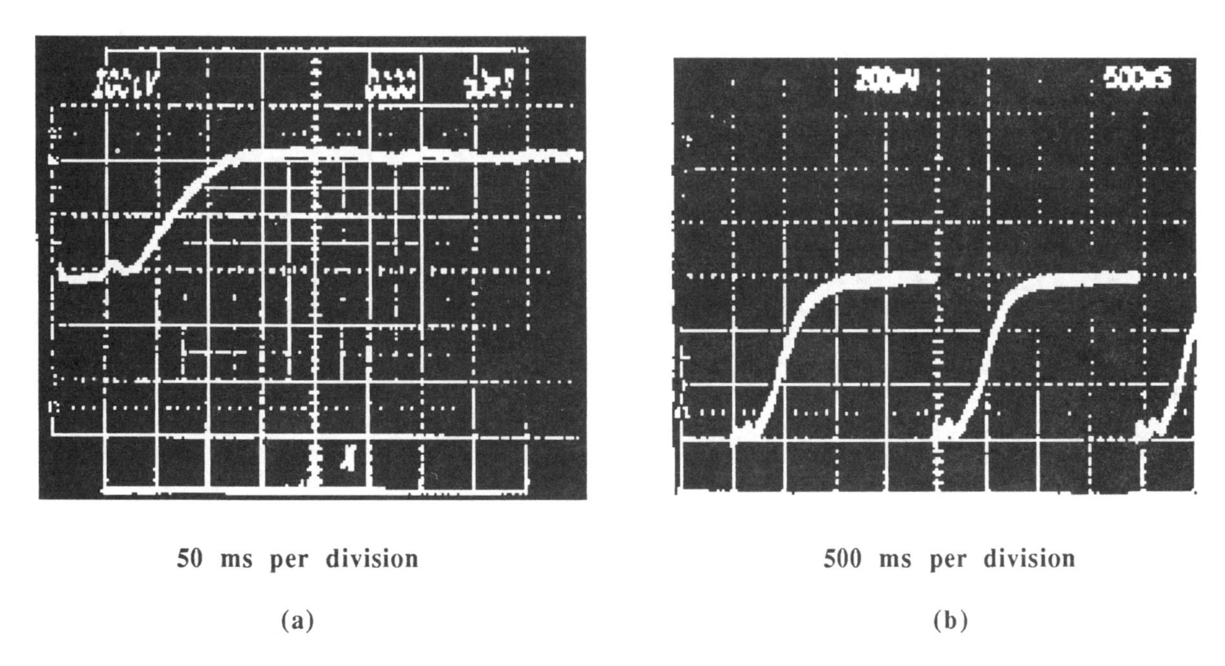

50 ms per division **500 ms per division**

(a) (b)

Figure 7. Thermal relaxation time measurements in (a) 1500 psig deuterium and (b) 1000 psig methane.

CONCLUSION

The Hughes Electro-Optical Tracking Sensor provides an Eyesafe laser rangefinding capability with data update 15 times a second to determine range and range-rate of uncooperative targets at distances of 10 kilometers or more. This fully militarized sensor system has been made possible by years of development in the technologies of high pulse rate Stimulated Raman Scattering lasers and InGaAs Avalanche Photodiode receivers. Hughes is continuing development of eyesafe laser systems by applying the same Raman cell technology to a Backward Raman cell configuration, which has demonstrated a factor of two improvement in beam quality. Preliminary experiments using Deuterium gas as the Raman medium demonstrates that 5 hertz, 20 milliJoule operation is achievable without circulating the gas in the cell. The goal of this effort is to develop a 10 to 20 hertz eyesafe laser with an uncirculated Raman cell.

ACKNOWLEDGEMENTS

We would like to recognize the team of engineers, lead by J. A. Bernal, who designed, fabricated, tested and delivered the first EOTS laser system during a short seven months in 1990. Key contributions were made by: R. A. Bork, H. L. Edwards, M. E. Ehritz, K. D. Goodin, D. E. Maguire, R. W. Manker, J. K. Miyamoto, M. Palombo, and C. N. Sakamoto.

REFERENCES

1. E. Gregor, D. E. Nieuwsma and R. D. Stultz, "20 Hz Eyesafe Laser Rangefinder for Air Defense," *Proc. SPIE Int. Soc. Opt. Eng.*, vol. 1207, p 124, 1990.

2. United States Environmental Hygiene Agency, *Nonionizing Radiation Protection Study No. 25-42-0332-90, Potential Eye and Skin Hazards with the Hughes "Eye-Safe" Air Defense Laser*, Aberdeen Proving Ground, MD, 1990.

3. B. W. Kennedy, B. A. Locke, W. Klimek and R. Laughman, *Large-Area and Self-Screening Smokes and Obscurants (Smoke Week XII) Quick Look Report*, Science and Technology Corporation Technical Report 4050, August 1990.

4. Parazzo, Buchman, and Stultz, "Numerical and Experimental Investigation of a Stimulated Raman Half Resonator," *IEEE J. Quantum Electron.*, vol. 24, pp. 872-880, June 1988.

5. H. W. Bruesselbach and D. R. Dewhirst, *Single Mirror Integrel Raman Laser*, U. S. Patent No. 4,821,272, April 1989.

6. B. A. Locke, W. M. Farmer, B. W. Kennedy and W. Klimek, *Joint United States-Canadian Obscuration Analysis for Smokes in Snow (Smoke Week XI) Quick Look Report*, Science and Technology Corporation Technical Report 3052, June 1989.

7. J. C. White, *Tunable Lasers*, Springer-Verlag, p. 116, 1987.

8. R. B. Lopert, "Measured Stimulated Raman Gain in Methane", Ph.D. Thesis, Univ. California Davis, 1983.

9. J. Ottusch and D. Rockwell, "Measurement of Raman Gain Coefficients of Hydrogen, Deuterium, and Methane", *IEEE J. Quantum Electron.*, vol. 24, no. 10, p. 2076, Oct. 1988.

10. R. W. Hellwarth, "Phase Conjugation by Stimulated Backscattering", Ch. 7 of *Optical Phase Conjugation*, edited by R. Fisher, Academic Press, New York, 1983.

11. B. N. Perry et al., "Stimulated Raman Scattering with a tightly focused pump beam", Opt. Lett., vol. 10, no. 3, p. 145, March 1985.

12. M. J. Shaw et al., "High-Power Forward Raman Amplifiers Employing Low-Pressure Gases in Light Guides II. Experiments", *J. Opt. Soc. Am. B*, vol. 3, no. 10, Oct. 1986.

13. M. Born and E. Wolf, *Principles of Optics*, Pergamon Press, New York, 1975.

Eyesafe diode laser rangefinder technology

A. Perger, J. Metz, J. Tiedeke, E. Rille

Leica Heergrugg AG
Heerbrugg, Switzerland

ABSTRACT

An eyesafe laser rangefinder technology based on repetitively pulsed diode lasers in the wavelength range between 800 and 900 nm has been developed. Single pulse energies and pulse train mean power have been calculated to be in accordance with eye safety regulations and to provide an optimum signal-to-noise ratio. Applications for this technology are short and long distance range finders which work on non cooperative targets.

1. INTRODUCTION

Laser rangefinders based on single pulse high power crystal lasers such as ruby or Nd:YAG are well known for more than twenty years. These lasers make use of pulses with a peak power of some hundred kilowatts up to some megawatts exceeding the limits of eye safety typically by a factor of 1000. So the use of these rangefinders is restricted to applications where exposure of human eye can be definitely avoided.

During the last years much effort has been made to develop eyesafe rangefinders based on the same principle as above using a single pulse of very high intensity, but at a wavelength which is not dangerous for the human eye. With lasers of this kind range measurement capabilities of 10 km or more have been obtained. The main disadvantage of these rangefinders is their complexity, poor efficiency and reliability, and cost.

Diode laser rangefinders available at present show the disadvantage of being restricted to ranges of a few hundred meters if the distance to non cooperative targets has to be measured. Their capability of measurement repetition rates in the kilohertz region has so far only been utilized for increasing their range accuracy.

A paper on gallium-arsenide eyesafe laser rangefinders [1] presented a technique which makes use of the high diode laser repetition rates for increasing the distance measurement capability. Improvements of this technique have brought about range capabilities in the kilometer region obtained at output powers not exceeding class 1 limits according to eyesafety regulations of IEC as well as of FDA.

2. MEASUREMENT PRINCIPLE

Common crystal laser rangefinders are based on the principle of the time-of-flight measurement. A very intensive light pulse, the duration of which is commonly of the same order of magnitude as the time resolution of the system, is transmitted to the target. In case of a noncooperative target only a very small amount of the transmitted power can be received as an echo signal. This optical signal is converted to a current pulse by a photodiode, amplified by a preamplifier unit and is then used for triggering a comparator. From the time delay between transmitted and received pulse the distance is calculated.

The performance of such a rangefinder is not only dependent on the emitted power but also on the sensitivity of the receiver, represented by the power capable of triggering the comparator. The trigger

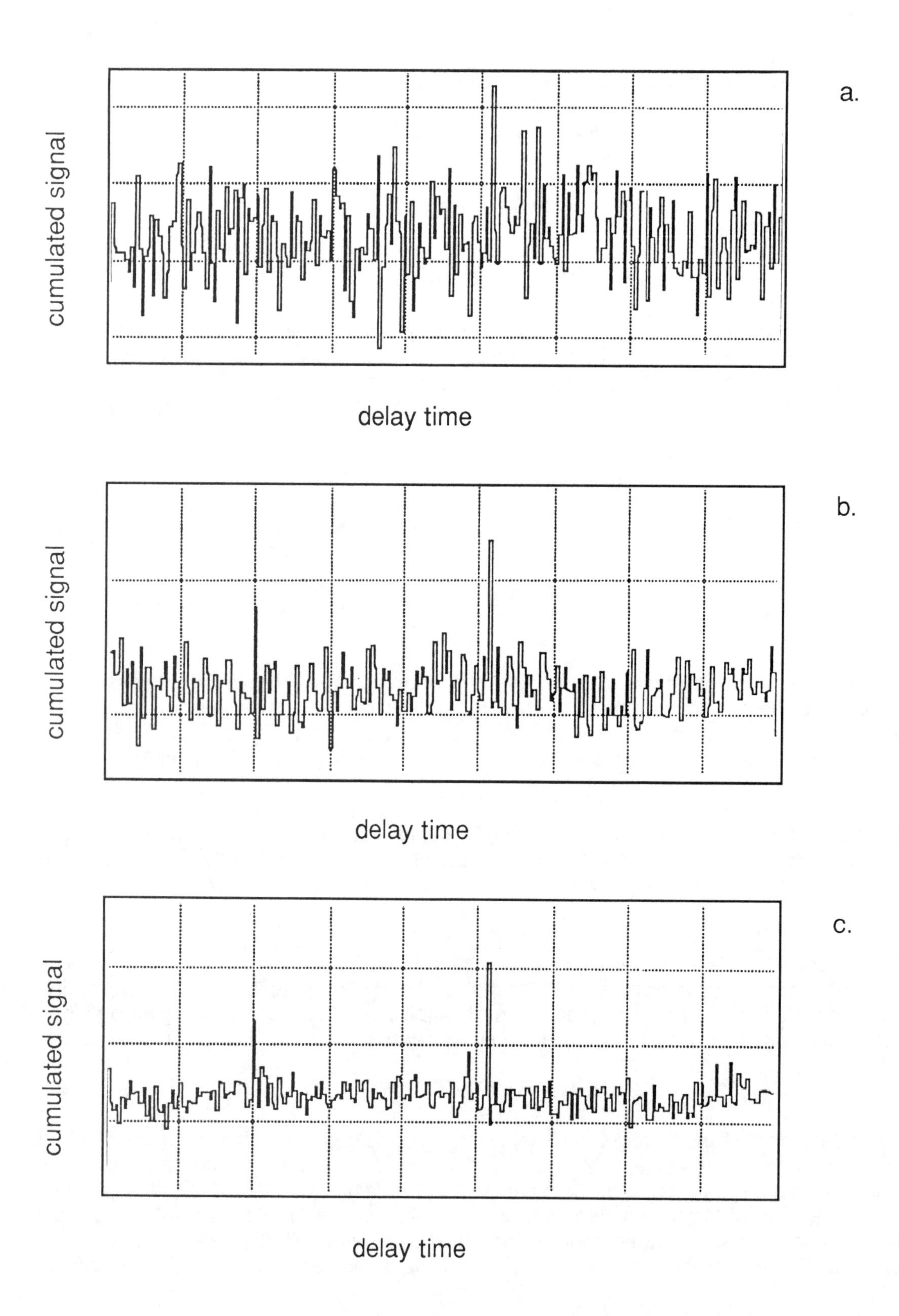

Figure 1: Improvement of SNR by increasing number of pulses

level of the comparator has to be well above the noise caused by background illumination, inherent photodiode noise and preamplifier noise.

The measurement principle developed and patented by LEICA HEERBRUGG AG was described in more detail elsewhere [1]. The corresponding rangefinder consists in principle of the same main functional units as conventional rangefinders. Because of the need of optimized sensitivity more effort has been made to reduce the noise of the preamplifier, to control APD gain and to reduce background illumination. The laser driver is provided with the capability of repetitive pulsing. The main differences are being obtained in the signal processing unit.

After emission of a laser pulse the receiver signal is sampled by an ADC and stored in a RAM forming a histogram of sampling values versus time delay. Because of the low peak power of the emitted pulses (typically a few watts) the echo signal is far below noise. If further pulses are transmitted and the information of the corresponding histograms is combined with the first one, the echo signal grows linearly with the number of emitted pulses, N, whereas the standard deviation of the noise grows proportional to the square root of N. So the signal-to-noise ratio increases with $N^{1/2}$. Figure 1 demonstrates the SNR improvement by the number of laser pulses. Figure 1a shows a histogram yielded by 1000 pulses. Figures 1b and 1c show the corresponding histograms for 3000 respectively 10000 laserpulses at the same conditions. By advanced post-processing of the histogram further improvements of the SNR can be obtained as well as distance measurement accuracies far beyond the value corresponding to the sampling rate.

3. EYESAFETY CONSIDERATIONS

Considering the potentials of this unique measurement technique the question for the limits of its capability arises. Dimensions of the receiver optics are commonly limited by the demand for a compact, light weight system. Sensitivity of the receiver is limited by the preamplifier noise and by background illumination. Improvements in this field are restricted by physical limits as well as by economical considerations.

As previously described, the two main parameters which influence the signal-to-noise ratio, are the pulse energy and the number of pulses. Increase of the values of these two parameters is limited mainly by eye safety considerations and by laser power capability besides some further aspects like duration of measurement, size of transmitter optics etc.

At LEICA HEERBRUGG AG different types of rangefinders are being developed. They can roughly be devided into two groups: rangefinders for short distances and rangefinders for large distances. All these rangefinders are subject to the regulations of class 1 of IEC as well as FDA. In both cases pulse energy and repetition rate for best performance and according to safety regulations have to be found out, taking into account the limits of the laser specification. The principles for optimization are the same for both groups of applications. However there are differences in the results because different values for transmitter aperture, beam divergence and some other parameters are being used. In the following the procedure of optimization is discribed for the case of large distance rangefinding.

The Accessible Emission Limits (AEL) according to IEC for a single pulse are

$$10^5 t^{0.33} C_4 \, J m^{-2} sr^{-1}, \qquad (1)$$

or

$$2 \cdot 10^{-7} C_4 \, J. \qquad (2)$$

In case of a pulse train the above AEL for the single pulse is reduced by $N^{-1/4}$. The FDA limits for a single pulse are

$$10 k_1 k_2 t^{1/3} J cm^{-2} sr^{-1}, \tag{3}$$

or

$$2 \cdot 10^{-7} k_1 k_2 J, \tag{4}$$

without further limits for the single pulse energy within a pulse train. Useful values for the above parameters are

t	50 ns,
C_4	1.95 (λ = 845 nm),
k_1	1.91 (λ = 845 nm),
k_2	1,

aperture diameter	40 mm,
measurement angle	10^{-5} sr.

This leads to AELs of 191 W (IEC) and 7.64 W (FDA) for a single pulse.

In case of a measurement duration of 100 ms the limits for the total energy of the pulse train are according to IEC

$$10^5 t^{0.33} C_4 J m^{-2} sr^{-1}, \tag{5}$$

or

$$7 \cdot 10^{-4} t^{0.75} C_4 J, \tag{6}$$

and according to FDA

$$10 k_1 k_2 t^{1/3} J cm^{-2} sr^{-1}, \tag{7}$$

or

$$7 \cdot 10^{-4} k_1 k_2 t^{3/4} J. \tag{8}$$

This leads to AELs of 1.15 mJ (IEC) and 1.11 mJ (FDA) for the pulse train.

Fig. 2 shows the limits for peak power versus number of pulses according to the different regulations. The "safe operation area" gives the range of power which is below all class 1 limits.

The next step is to find out the point of maximum rangefinder performance within this "safe operation area". As mentioned above the signal-to-noise ratio is proportional to peak power P_{peak} as well as to the square root of the number of pulses N. Since accessible range and accuracy are dependent

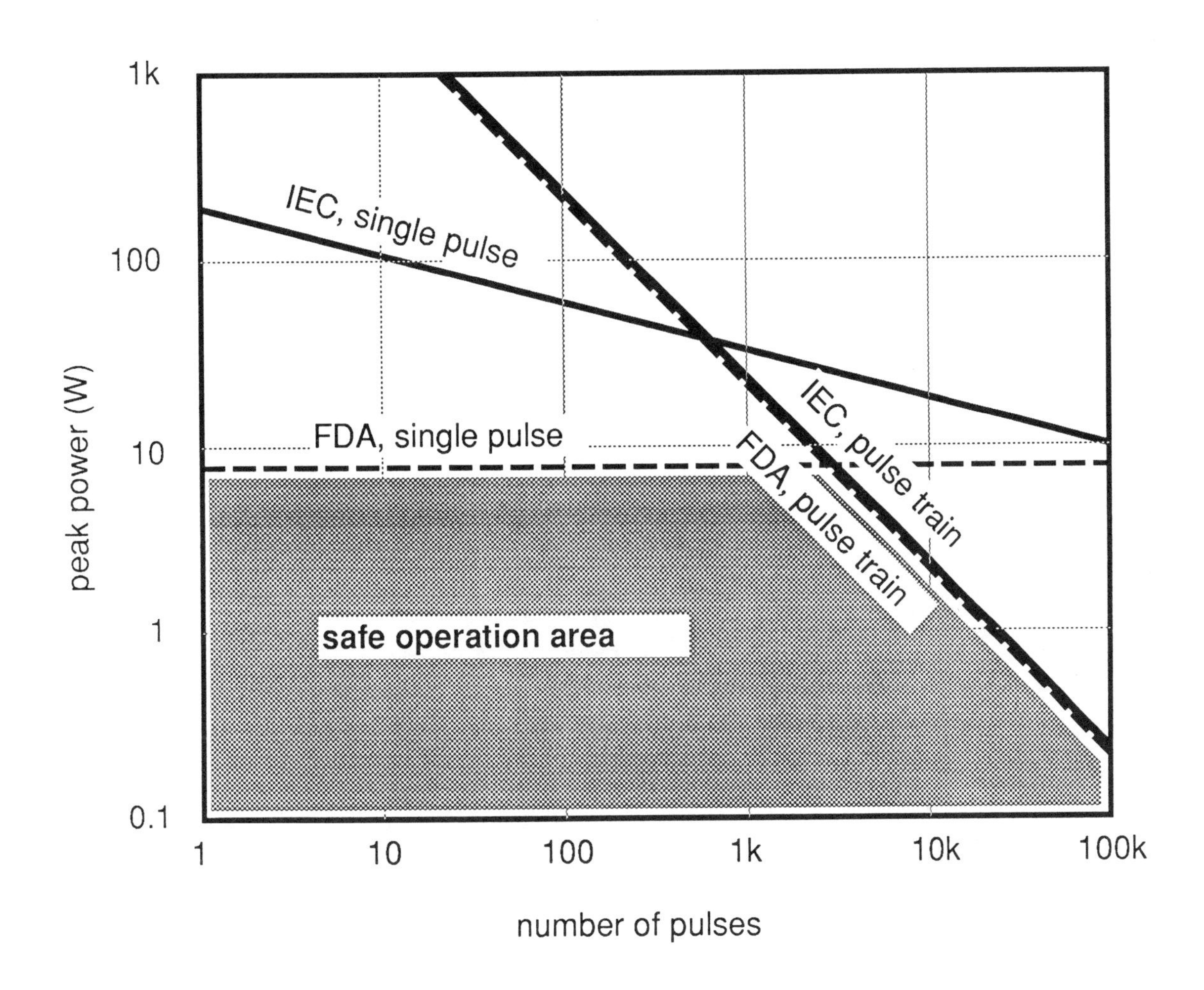

Figure 2: Area of eyesafe laser operation for 50 ns pulses at an aperture of 40 mm and wavelengths beyond 845 nm.

on the SNR, peak power and number of pulses are the key parameters for rangefinder performance improvements.

In figure 3 besides the "safe operation area" curves of constant rangefinder performance are shown. Because of the double logarithmic scale used in figure 3 they are straight lines These lines represent constant values for $P_{peak} \cdot N^{1/2}$. The "point of best eyesafe performance" can easily be found. It corresponds to a peak power of 7.64 W and a number of 2900 pulses resulting in a pulse repetition rate of 29 kHz.

4. RANGEFINDER SYSTEM DESIGN

The results of the eye safety considerations fit well into the capabilities of diode lasers and signal processing. System design starts for example with the choice of the transmitter lens diameter which is determined by the overall size of the rangefinder instrument. The focal length of the transmitter lens is then limited by the f-number in order to obtain a sufficient fraction of the total laser power being emitted. The permissible size of the emitting area of the laser diode can be calculated taking into account the requirements for beam divergence which are commonly in the range between 1 and 2 mrad for long distance measurement applications.

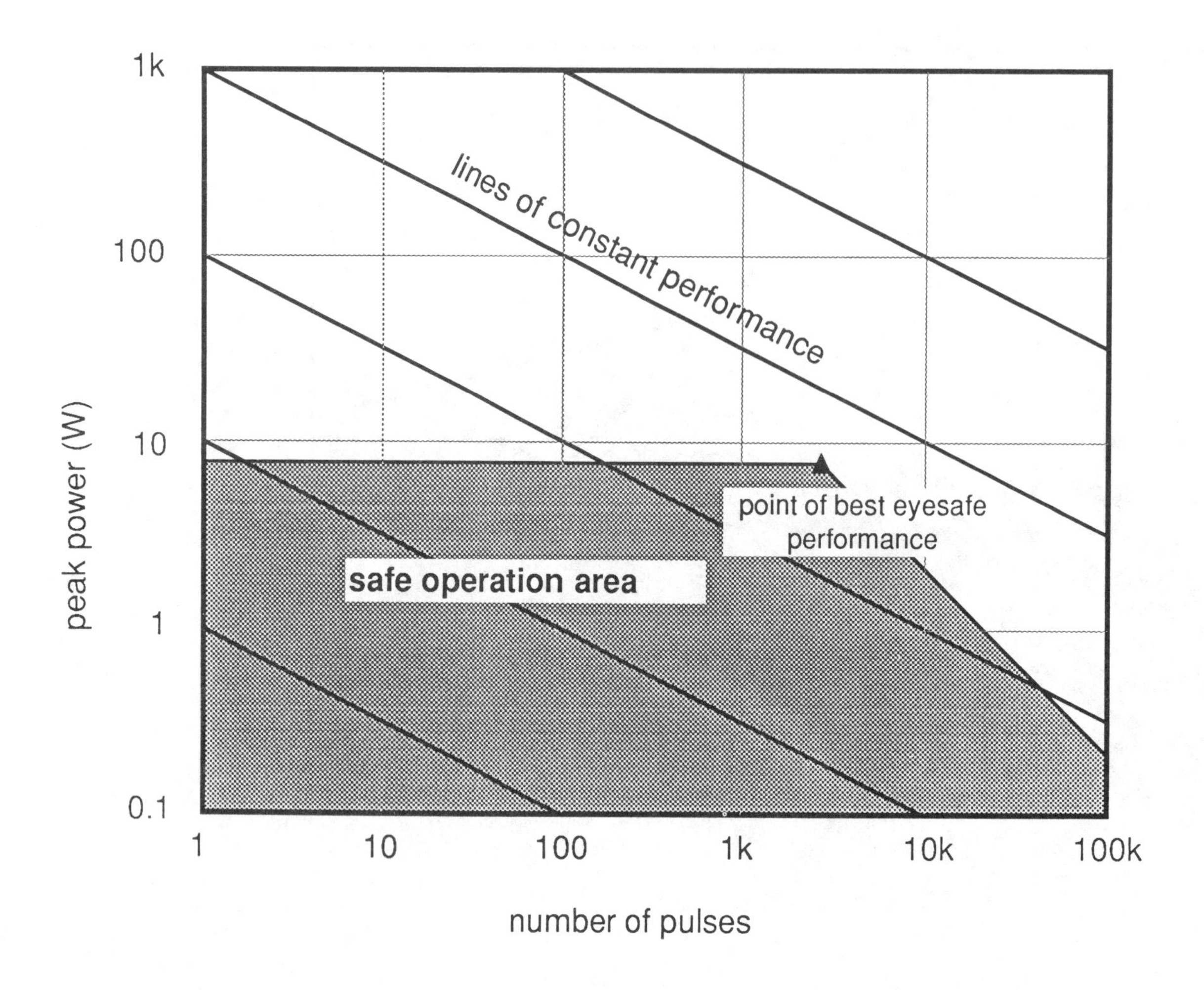

Figure 3: Safe operation area and lines of constant rangefinder performance; point of best eyesafe performance

Available pulse diode lasers with the above restrictions on emitting area size commonly do not reach the eyesafety limits of peak power given in the previous section. Since the eyesafety limits of peak power and pulse train energy are given by expressions for the power density respectively energy density and the density is somewhat elevated in the centre of the transmitter lens it is advisable to operate at lower levels of single pulse energy as well as of pulse train energy. Operating the lasers at levels about 50 percent of the above safety limits gives one the certainty to take the power density elevation into account as well as the specification of the laser in order to obtain sufficient lifetime. It also provides some safety margins concerning temperature and lifetime drifts of laser power and some additional influences.

Other aspects of system design as well as target properties and influences of the atmosphere have been reported earlier [1]. Some progress has been obtained also in other fields influencing rangefinder performance, for example concerning the avalanche photodiode control.

5. RANGEFINDER PERFORMANCE

5.1. Theoretical predictions

The performance of a rangefinder can be expressed by the minimum yield the rangefinder can cope with. The yield is defined as the power density at the location of the receiver, devided by the transmitted power. It is independent on the rangefinder properties, such as receiver aperture etc.

In case of a target perpendicular to the laser beam, larger than the spot size of the laser and with a Lambertian reflectance characteristic, the yield Y can be calculated as follows

$$Y = A \cdot T \cdot r^{-2} \pi^{-1}, \qquad (9)$$

where A is the target reflectivity, T the transmittance of the atmosphere and r the distance to the target.

The minimum yield sufficient for reliable rangefinding is given by

$$Y = P_R \cdot P_{peak}^{-1}, \qquad (10)$$

in the case of single shot rangefinders and by

$$Y = P_R \cdot P_{peak}^{-1} N^{-1/2}, \qquad (11)$$

in the case of repetitively pulsed laser rangefinders where P_R is the minimum power density required by the receiver to achieve a certain detection probability. Typical values for P_R are in the order of magnitude of 1 μWm^{-2}.

Figure 4 shows the yield at 865 nm for target reflectivities of 0.2 and 0.4 and for infinite visibility as well as threefold visibility. Threefold visibility means that objects in a distance of three times the target distance can just be perceived.

As can be seen in figure 4, ranges up to 4000 m can be measured by the repetitively pulsed rangefinder at good conditions. If target and visibility conditions are poor, ranges beyond 1000 m are still being obtained. A single pulse laser rangefinder of the same type (concerning optic dimensions, peak power etc.) is only capable of ranges of about 500 meters under good conditions.

5.2. Experimental results

Performance measurements on a LEICA diode laser rangefinder have been made at different conditions for visibility and target properties and at different ranges in order to investigate the relation between theoretical and experimental results. The signal-to-noise ratios and distance accuracies obtained at these tests were in good agreement with the predictions derived from the above calculations.

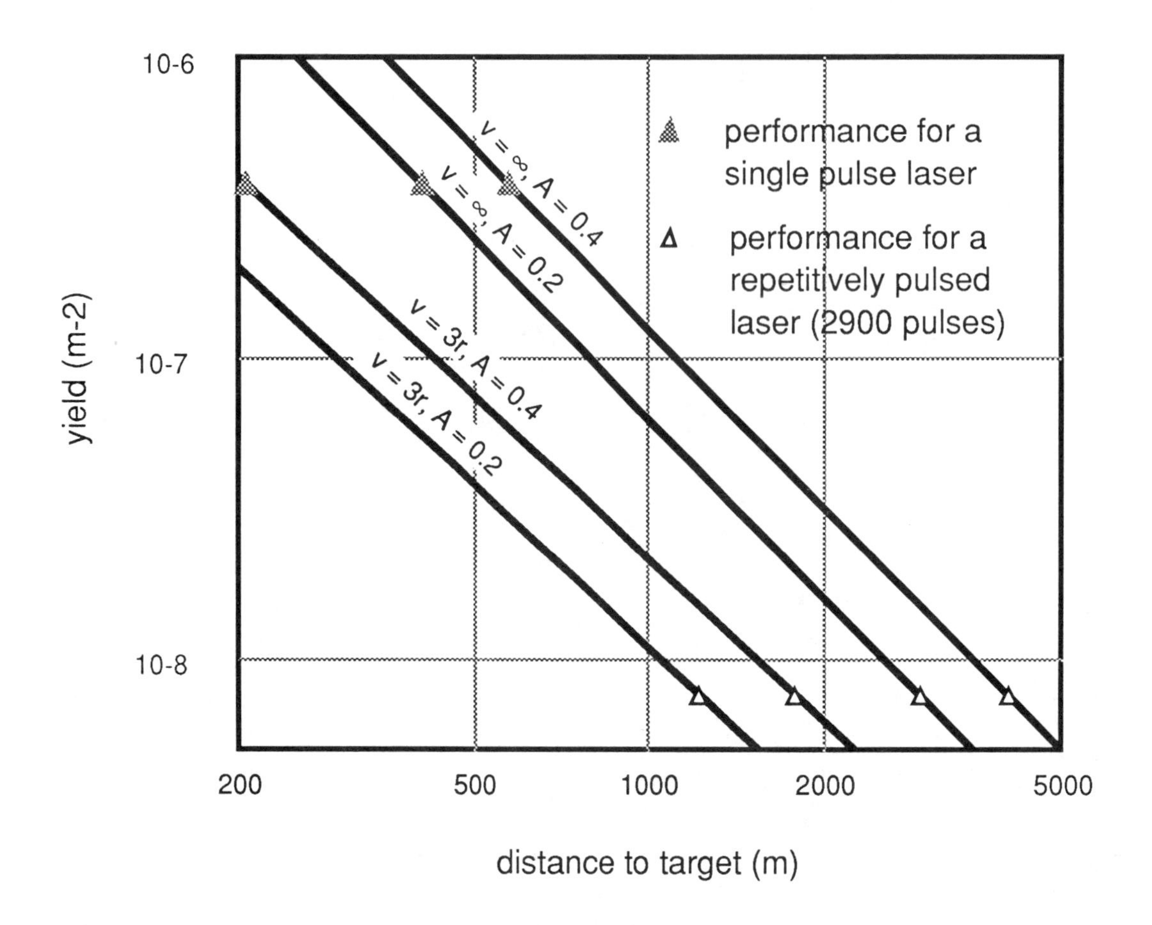

Figure 4: Yield for different target and visibility conditions at 865 nm

6. APPLICATIONS

6.1. Short range

In the next few years diode laser rangefinders will become an important accessory for cars and trucks in order to enhance safety as well as comfort and continuity of traffic. Application in automotive industry is characterized by high quantities, high demand on reliability and low price targets.

Near future applications for those rangefinders will be convoy driving, where the rangefinder determines and simply displays the minimum safety distance. A next step will be intelligent cruise control, where a common cruise control will get the information to reduce speed out of the rangefinder if the minimum safety distance has been undercut. A higher step will be autonomous intelligent cruise control, where the rangefinding sensor is adapted to throttle and brakes for distance control. Several national and international R&D Programms are working on these topics. Within the scope of the PROMETHEUS research program the European Automotive industry is running first field tests using cars equipped with the LEICA Multisegment Distance Measurement System (MSAR) [2].

New laws for German highways saying that heavy trucks have to keep a minimum distance, created a demand for the first product. This sensor will simply help the driver to keep this minimum distance.

The technical characteristics of automotive distancers are ranges up to 200 m, high measurement rates (some ten to some hundred distance measurements per second) for determination of relative speed, accuracies in the decimeter range, parrallel ranging in several segments (or tracking capabilities), high angular coverage compared with long distance rangefinders resulting in larger amounts of background illumination and high dynamic ranges.

6.2. Long range

Distance measurement capabilities in the kilometer range are interesting for a number of applications. The additional feature of rangefinding to monocular and binocular observation instruments (day as well as night vision) raises their utility considerably. Handheld instruments can be used up to ranges of 1200 to 1500 m and are especially suited for the needs of hunters and yacht owners and, in the military domain, for infantry applications.

Rangefinders used beyond 1500 m have to be operated on tripods or other mounts. The main applications are military (i.e. rangefinders for anti tank weapon systems), maritime (distance measurement to buoys or to the shore) and avionic purposes if the visibility is high enough (as it is commonly the case in aerial surveying where an accurate measurement of the distance to ground is important for scale determination).

7. OUTLOOK

The above technology offers potentials for increase in accuracy up to the centimeter to decimeter range if sufficient SNR can be obtained. Higher accuracies are not being aimed at because this field is well covered by geodetical instruments. It also makes no sense in most cases to measure to noncooperative targets like trees, houses, cars etc. with millimeter accuracy.

An increase in range as well as in accuracy is bound to SNR improvements, as mentioned above. SNR can either be improved by increasing receiver sensitivity or transmitter energy. Increase of transmitter energy can either be realized by raising the measurement time resulting in a higher number of pulses or by enlargement of transmitter optics allowing higher output powers being emitted without exceeding eyesafety limits.

Increasing the receiver optics diameter is likely to be the most effective way for improvements on receiver sensitivity. Modifications of APD and preamplifier will bring about only small amounts of SNR improvements. The most serious source of noise is background illumination. By narrowing the optical filter bandwidth the influence of background light can be attenuated.

Based on the present technology a considerable increase in performance is possible but may result in modifications which influence system dimensions, measurement duration and other properties which bring about immediate disadvantages to the user.

8. REFERENCES

1. R. Brun, "Gallium-Arsenide Eyesafe Laser Rangefinder," *Laser Safety, Eyesafe Laser Systems, and Laser Eye Protection*, Penelope K. Galoff, David H. Sliney, Editors, Proc. SPIE 1207, pp. 172 - 180 (1990).

2. J. Tiedeke, P. Schabel and E. Rille, "Vehicle Distance Sensor Using a Segmented IR Laser Beam," *40 th IEEE Vehicular Technology Conference,* pp. 107 - 112, IEEE, New York 1990.

SIRE: An Eyesafe Laser Rangefinder for Armored Vehicle Fire Control Systems

H. S. Keeter, G. A. Gudmundson, and M. A. Woodall

Optic-Electronic Corp.
Imo Industries, Inc.
11545 Pagemill Road, Dallas TX 75243

ABSTRACT

The Sight Integrated Ranging Equipment (SIRE) incorporates an eyesafe laser rangefinder into the M-36 periscope used in tactical armored vehicles, such as the Commando Stingray light tank. The SIRE unit provides crucial range data simultaneously to the gunner and fire control computer. This capability greatly reduces "time-to-fire", improves first-round hit probability, and increases the overall effectiveness of the vehicle under actual and simulated battlefield conditions. The SIRE can provide target range up to 10-km, with an accuracy of 10-meters. The key advantage of the SIRE over similar laser rangefinder systems is that it uses erbium:glass as the active lasing medium. With a nominal output wavelength of 1.54-microns, the SIRE can produce sufficient peak power to penetrate long atmospheric paths (even in the presence of obscurants), while remaining completely eyesafe under all operating conditions. The SIRE is the first eyesafe vehicle-based system to combine this level of accuracy, maximum range capability, and fire control interface. It simultaneously improves the accuracy and confidence of the operator, and eliminates the ocular hazard issues typically encountered with laser rangefinder devices.

1. INTRODUCTION

The use of laser rangefinders in fire control systems for armored vehicles has greatly increased the accuracy and effectiveness of the vehicle in battlefield conditions. The laser range data fed to the fire control computer causes a reduction in time to fire as well as increasing first round hit capabilities. Due to the concern for safety and the need for realistic training tactics, the use of an eyesafe laser becomes highly desirable. When this is combined with the recent changes in modern combat scenarios of urban warfare, the need for eyesafety becomes mandatory.

Optic-Electronic Corp. has addressed this requirement by developing the Sight Integrated Ranging Equipment (SIRE) system. This system consists of an erbium glass laser rangefinder, a daylight optical channel, a night vision optical channel, and a CRT projection unit on a periscope assembly. SIRE is currently integrated on the Cadillac Gage - Commando Stingray Light Tank fielded by the Thailand Army.

This paper will discuss the function of the fire control systems in armored vehicles, with emphasis on the integration of eyesafe laser rangefinders. Laser specifications will be addressed in relation to effectively doing the job of ranging to the target in battlefield conditions. Data on the performance of the system in various environmental conditions and common battlefield obscurants will also be presented.

2. FIRE CONTROL SYSTEM REQUIREMENTS

Fire Control Systems are used on armored vehicles to control the weapons by giving automatic aiming information to the gun. This is done by using the range information obtained from the range finder while viewing the target. The range data is processed by the fire control computer and information on the elevation and aim point are calculated. This information is then used to move the weapon to the proper orientation and the weapon can then be fired.

The purpose of a fire control system is to reduce the time to fire and increase the first round hit probability. Hence, increasing the accuracy of the weapon system. To ensure this, the range data should be accurate to within 10 meters. The ability to obtain this accurate data in a timely manner necessitates the use of a laser rangefinder (LRF). The LRF must be able to obtain the data at a sufficient rate to accurately and rapidly locate targets. Generally a firing rate of one pulse per 3 to 10 seconds is all that is required. The fire control system must also be able to handle several different types of weapons and/or ammunition. This is to allow for the different velocities and ranges for each weapon and ammunition type. To be able to handle this requirement effectively, a fire control computer is needed to process the information.

The fire control system must be useful under a variety of battlefield conditions. This demands that the system allow the target to be seen at a reasonable distance, on the order of 5-kilometers or greater. In general, military systems require 7-power magnification to allow recognition at a these distances while still allowing a relatively large field of view. This capability is also required at night, and this can be meet with either an image intensifier tube or a FLIR integrated into the system.

Some of the more adverse battlefield conditions that the fire control system must be able to function under are weather conditions such as rain, fog, and snow, as well as both natural and man made obscurants like dust, sand, and smoke. These needs dictate that the power of the laser rangefinder be sufficient to penetrate the obscurants and weather. They also indicate the need for a wavelength compatible with the conditions. This is a difficult situation because there is no one wavelength that is good under all of the above conditions. Thus trade offs must be made to select the wavelength that will give the greatest benefits.

The three common wavelengths used for laser rangefinder are 1.06-μm, 1.54-μm, and 10.6-μm. The two shorter wavelengths are very comparable in performance with the 1.54-μm being slightly better for penetrating most of the obscurants. The 10.6-μm has good performance through the smokes, but does not perform as well through rain, fog, and snow as the shorter wavelengths due to water absorption.

In order for a fire control system to be effective, the user must be able to train frequently so he is familiar and confident with the system. Because of new regulations on laser use throughout the world, it has become increasingly difficult for the user to train with a laser rangefinder that is not completely eyesafe. Training missions with non-eyesafe systems are now restricted to designated laser ranges, and due to the limited number of these areas, time on them becomes a premium. Thus, the user only gets to train with the laser rangefinder once or twice a year and thus, is not sufficiently confident and/or well versed in using the laser rangefinder in conjunction with the fire control system. To further complicate matters, recent global political conditions have changed the battlefield scenarios to urban and "friendly" populated areas. This means that the likelihood of firing the laser on "friendlies" or ones own troops greatly increases. The way around this dilemma is to also require the use of an eyesafe laser in the battlefield.

TABLE 1

Maximum Permissible Exposure Limits
for Selected Laser Wavelengths[(1)]

Type of Laser	Wavelength	Exposure	Exposure Limit for Intrabeam Viewing
Ruby	694 nm	1 nS-18 uS	0.5 uJ/cm^2
Nd:YAG	1064 nm	1 nS-50 uS	5.0 uJ/cm^2
Er:Glass	1535 nm	< 1 uS	1,000,000 uJ/cm^2
Er:YLF	1730 nm	1 nS-100 nS	10,000 uJ/cm^2
Ho:YLF	2060 nm	1 nS-100 nS	10,000 uJ/cm^2
CO_2	10600 nm	1 nS-100 nS	10,000 uJ/cm^2

As can be seen from the Maximum Permissible Exposure limits shown in Table 1, any laser can be considered eyesafe if the power level is sufficiently low. Several wavelengths (1064-nm and shorter) can be eliminated because the low energy at eyesafe levels will not allow ranging to the distances required. Other wavelengths (1730-nm and 2060-nm) are not practical for tactical military systems due to the relatively low maturity of the basic hardware, unacceptable design requirements (cryogenic coolers), or high technical risk. This leaves only two practical wavelengths for further consideration, namely 1.54-μm and 10.6-μm. Both of these wavelengths are currently being used in military systems.

Because of the need for covertness, provisions must be made in the fire control system to allow it to be used while the vehicle is off and to eliminate excess noise from the system. This dictates the use of batteries to power the components. The noise requirement discourages the use of components that require coolers, fans, and pumps. This makes the use of FLIR systems, as well as far infrared laser sources, less desirable. Covertness also indicates the need for wavelengths that are not easily detectable. This tends to eliminate the use of visible light. It also makes the Nd:YAG wavelength less acceptable because of the common use of laser threat warning devices sensitive at that wavelength. It makes the use of a CO_2 laser at 10.6-μm less desirable, because it is easily seen by the common FLIR systems in use. The 1.54-μm wavelength is undetectable by any fielded sensor system.

Another requirement is to protect the user and equipment from laser damage caused by incident light on the optics of the system. This is needed for a variety of reasons. The first is to protect the system from accidental exposure by ones own forces during both training and combat situations. The second is to protect against enemy lasers used for rangefinding, designation, and as offensive and defensive blinding weapons.

Lastly, but by no means the least important, cost is always a major consideration. The fire control system should be able to meet the above requirements while being cost effective. This creates many important trade offs that must be seriously considered. Should one go with the use of the expensive FLIR system and its increased performance or go with the less expensive, but more restricted night vision device? Should a high energy, higher rep-rate laser be used or a single shot low power one that costs less? These are but a few examples of the questions that cost raise.

Given the above requirements, and after weighing the trade offs, Optic-Electronic Corp. (OEC) has developed the Sight Integrated Ranging Equipment - SIRE.

3. SIGHT INTEGRATED RANGING EQUIPMENT (SIRE)

A photograph of the SIRE system incorporating an M-36 periscope is shown in Figure 1. SIRE is c also compatible with a variety of other periscope designs, including the MP-86 multi-purpose sight. Aside from the basic periscope, the major components of the SIRE system are an Eyesafe Laser Rangefinder (LRF unit), a Passive Night Vision Elbow (PNVE), and a CRT Projection unit mounted into a Head Assembly. The LRF is powered by, and interfaced to, a fire control computer via the Laser Power Unit (LPU). As an option the system can also have a Commanders Remote Display (CRD).

The Head Assembly supports the periscope in a M119E1 mount and connects to the turret ballistic drive via an adjustable input arm. It contains a unity sight with a horizontal field of view of 60 degrees, and a vertical field of view of 10 degrees. The head mirror can be rotated up to +60 degrees and -20 degrees in elevation. All the optical systems view through the head mirror and are thus linked to the turret ballistic drive and consequently to the main gun.

The Passive Night Vision Elbow is a 25-millimeter second generation Image Intensifier Tube (IIT) that allows use of the system at night. The optical channel of this device gives 7-power magnification and a 7-degree field of view. The PNVE contains a ballistic reticle that is boresighted to the LRF. The IIT is powered by 24 VDC which makes it compatible with vehicle power. It also contains rechargeable NiCad batteries to allow silent watch operation when the turret power is off.

The CRT projection unit interfaces with the fire control computer, the LRF and optionally the PNVE. It receives electrical power from the fire control computer. This unit projects an aim point, as calculated by the fire control

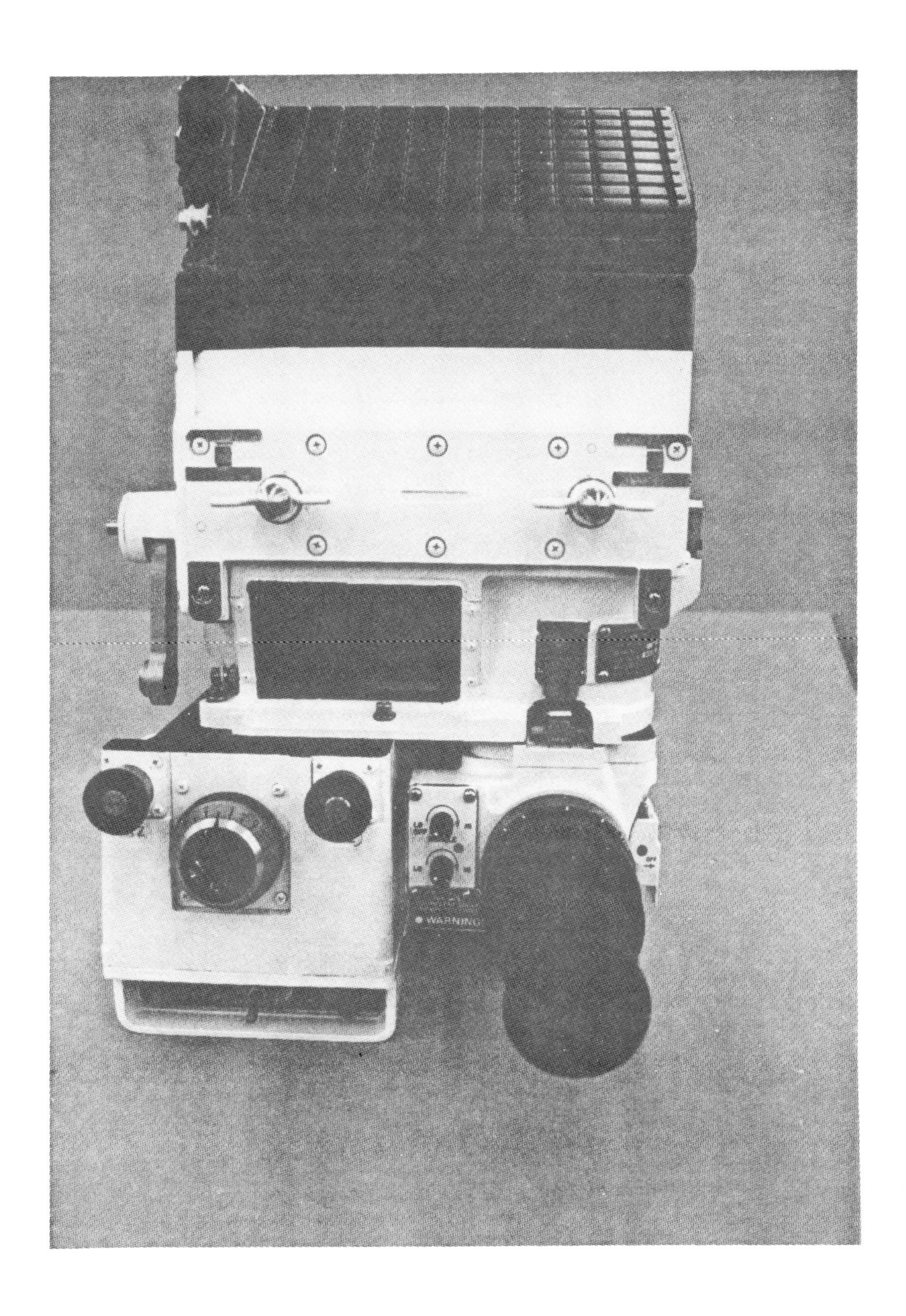

Figure 1 - The SIRE system uses an eyesafe laser to provide accurate range capability for a variety of armored vehicles.

computer, into the optical system. This aim point is then viewed in the eye piece of either the LRF or the PNVE. The aim point is then placed on the target, which effectively aligns the weapon for firing. This use of a projected aim point greatly increases the first hit capability of the vehicle.

The Laser Power Unit (LPU) is a separate module connected to the SIRE LRF assembly by shielded cables. The LPU contains the low voltage regulators and power conditioning circuitry, the high voltage power supply, and the computer interface circuitry. The LPU receives vehicle power through one of its connectors and provides the electromagnetic interference (EMI) protection, power conditioning, and stabilization to run and protect the LRF and computer interface circuitry. The LPU converts the low voltage into the high voltage the laser pulse forming network (PFN) needs to operate the laser transmitter.

This high voltage as well as the interface signals to and from the LRF are carried through a separate connector to the LRF. A third connector handles the interface with the fire control computer. This interface is essentially a strobed binary-coded decimal (BCD) transmission of the range data. The laser can be fired from the gunners control handle via the computer, or from a button on the LRF itself. The LPU also houses the rechargeable batteries and battery charger for the LRF. This allows the LRF to be operated in the silent watch mode.

One option on the system is the Commanders Remote Display. This is a small box that contains a 4-digit LED display so the commander may independently view the range calculated by the LRF. The commander also has the capability to fire the laser from the remote display.

4. LRF SUBSYSTEM

A block diagram of the LRF subsystem is shown in Figure 2. The major component of the SIRE system is the LRF. The SIRE LRF is a laser transmitter with coaxial sighting optics and receiver. The range counter and display electronics are included in this unit as well as the laser PFN. The power for operating this unit is provided via a cable from the Laser Power Unit (LPU).

An optical schematic of the SIRE system is provided in Figure 3. The sighting optics are a visual optical channel with 7-power magnification and a 50-mm aperture. The field of view of the visual system is 8-degrees. The reticle has crosshairs with mil graduation marks for ease of aiming. The laser transmitter and receiver are independently boresighted to the reticle. The eyepiece includes special laser protection features to reduce the chance of operator injury by reflected or direct laser beams. Mounted to the reticle is the range display. This display consists of four 7-segment LEDs to show the range, an LED for the ready to fire signal, and a LED for indicating when Multiple targets have been detected.

The laser transmitter is a Q-switched erbium glass laser operating at 1.535-μm. The output energy is typically 20-mJ, and the pulse width is a nominal 25-nS. The laser is capable of firing every 3-seconds for limited bursts, or once every 10-seconds indefinitely. The beam divergence is typically less than 0.6-mR at the 90% energy points. This low beam divergence allows for better target discrimination at longer distances due to the smaller beam size.

The receiver is an InGaAs pin photodiode with its associated electronic amplifiers on a single board. The receiver field of view is 1.5 mR full angle. This is at least twice the transmitted beam size and allows for maximum collection of the return laser energy from the target even with slight boresight errors. This receiver has overload protection to prevent damage to the electronics in case of a large amplitude return signal. It also includes a Time-Programmed Gain (TPG) circuit to ramp the gain in a close approximation to the $1/R^2$ signal strength relationship of the return signal from the target. This circuit reaches maximum gain at around the 2.5-km range time period. This allows maximum signal detection from targets that return small signals because of low reflectivity or being smaller than the beam.

The rangecounter circuitry is initialized when the outgoing laser pulse (t-zero) is detected by the receiver from the backscattered light from the common optics. The receiver processes the laser light into a digital signal that is

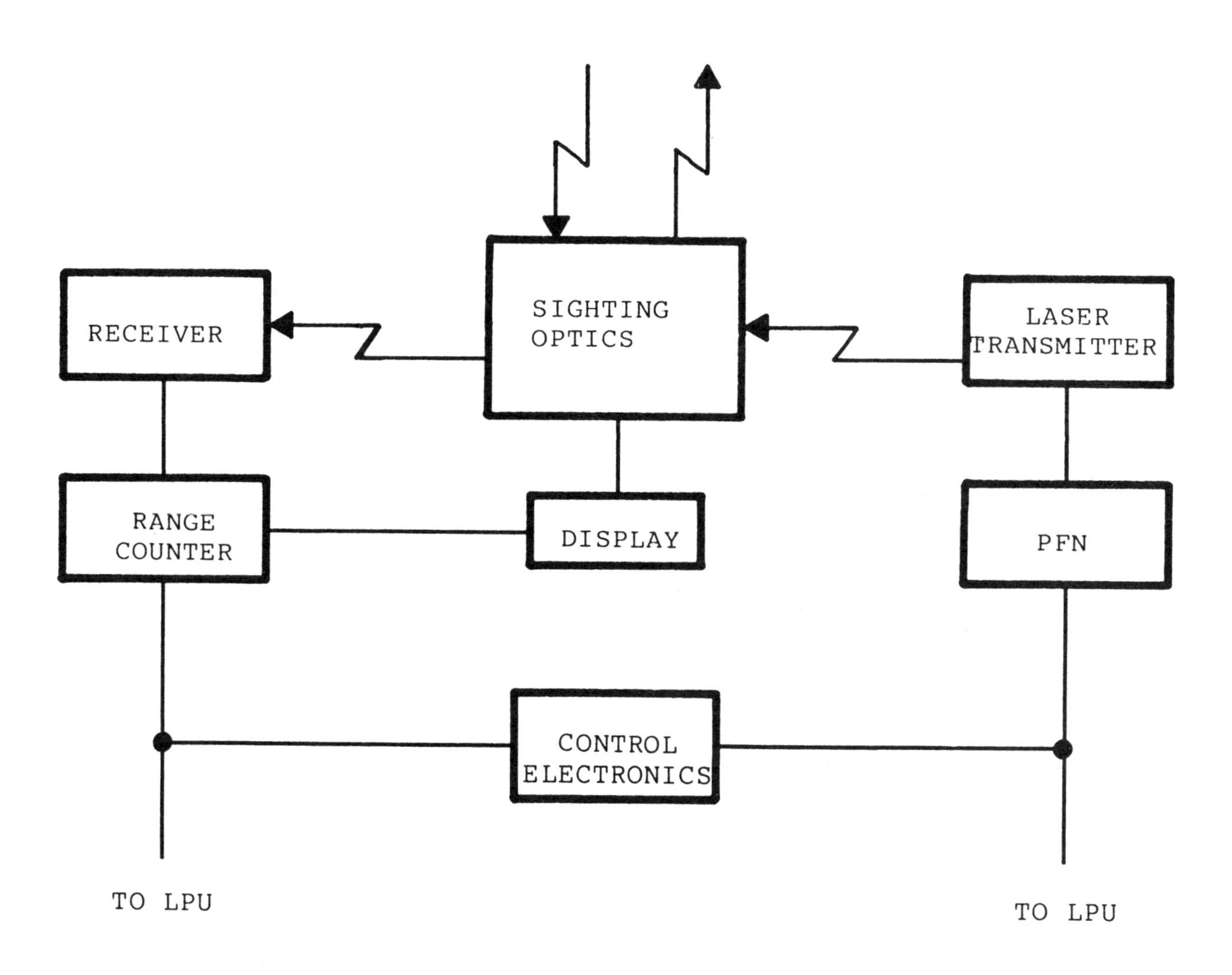

Figure 2 - Block diagram of SIRE laser rangefinder (LRF) subsystem.

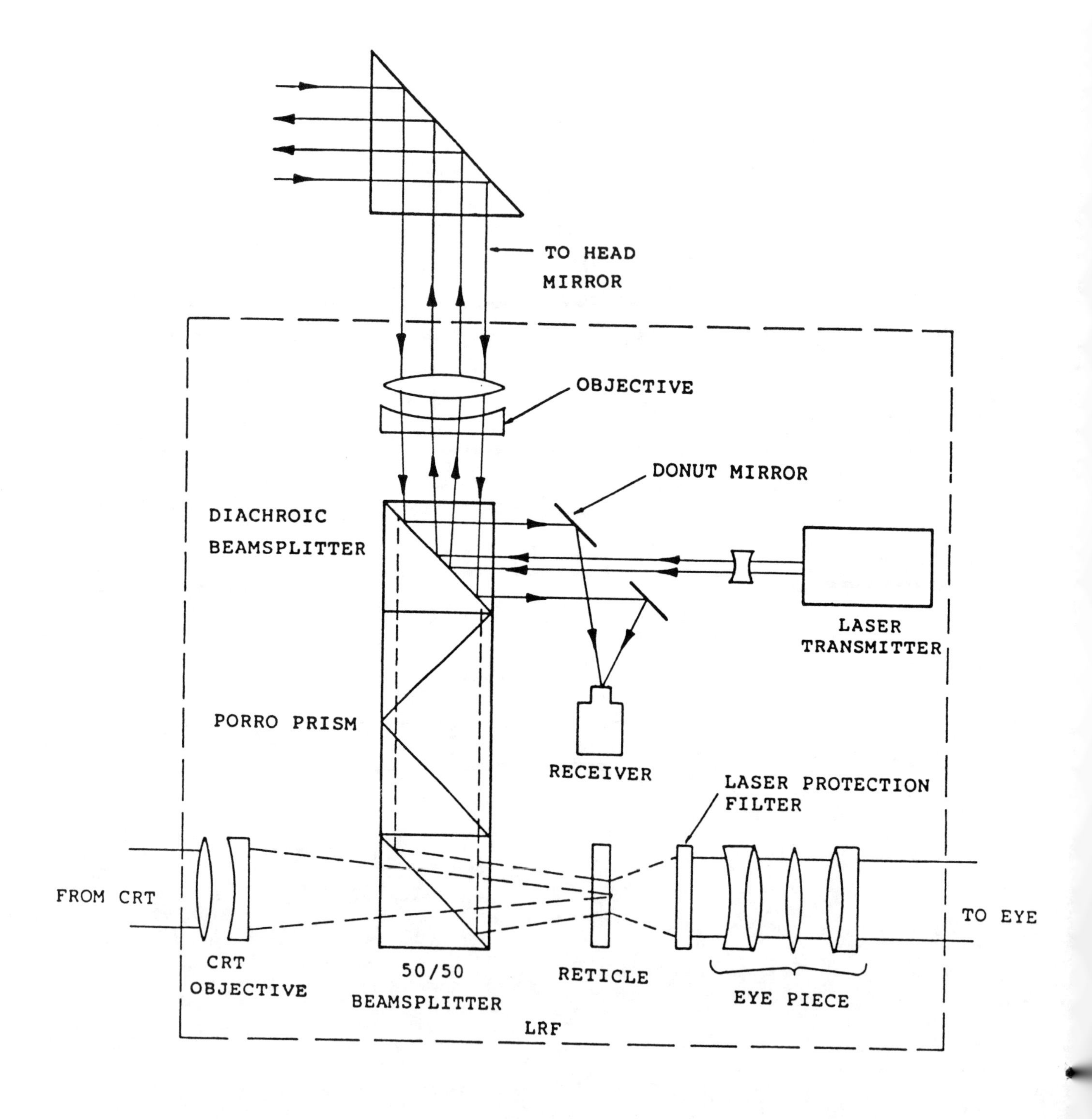

Figure 3 - The SIRE uses a common optical axis for the transmitter, sighting, and receiver paths.

transmitted to the rangecounter. The rangecounter counts the time between the t-zero pulse and when the return signal is received. This time is then processed into a range value that is sent to the display in the eyepiece. It is also sent, via the LPU, to the fire control computer and to the optional commanders remote display if it is in the system. Ranges below 200-meters are blocked out of the rangecounter electrically to prevent false targets from atmospheric backscatter and ground clutter. The maximum range that can be displayed by the rangecounter is 9995 meters. This results in an effective range gate of 200 to 9995-meters for the LRF which is more than adequate for the fire control system. Short ranges are not needed because the weapons are line-of-sight at this distance and, if a target is that close its too late to worry about the range anyway. The maximum range is well beyond the range of the weapons on the vehicle and further distances are of little interest.

The accuracy of the rangecounter, and thus the LRF, is 10-meters with a 5-meter resolution. The rangecounter also includes selectable first/last ranging logic. This feature is selected on the front panel of the LRF by the operator. It is used to discriminate between multiple targets. The "first" mode will display the range of the first target received. In "last" mode it will continually update the register with the last range received until the 9995-meter range gate is passed. It then displays the range of the last target from which a return signal is received. This circuit is of obvious use when ranging through a tree line or bushes to a target. During this situation one would operate in the last target mode. Likewise, when ranging to a small target first logic is the best choice because it prevent the laser beam spill over from giving incorrect ranges. The first/last logic can discriminate between targets with as little as 50-meters separation.

5. FIELD PERFORMANCE

More than 100 SIRE systems have been fielded in the Cadillac Gage - Commando Stingray Light Tank by the Thailand Army. These units have been in service for two years with few problems. The laser rangefinder, when coupled with a fire control computer, greatly increases the first round hit probability. It has also been found that the number of rounds and time to kill is significantly reduced. Thus, it is possible to deliver more rounds on target in a shorter time period. This is supported by data from field testing of the SIRE, and also confirmed by a similar system installed in a Bradley Fighting Vehicle (BFV) which participated in Operations Desert Shield and Desert Storm. During these tests it was found that the first hit probability was increased by 300%, and time to kill was reduced by 400%. Similar improvements in effectivity have been observed under various battlefield conditions, including night operations from a moving vehicle and against stationary, moving, and airborne targets.

In testing at OEC, and using a similar system at MICOM in Huntsville, Alabama, it has been found that the erbium laser rangefinder functions very well in light-to-medium rain and fog. At MICOM it was found that in light rain, the range to T-62 tanks at distances up to 5-km could be easily obtained. As the rain fall rate increases, performance was found to be roughly equal to the visibility. This has also been found to be true during testing at OEC in Dallas. The general rule of thumb is <u>if the target can be seen, it can be ranged to</u>.

OEC has used the SIRE LRF at the last two Smoke Week trials conducted by the US Army. Extensive testing was conducted to determine the effects of man made obscurants on the SIRE LRF[2]. These results are summarized in Table 2. It is obviously out of the scope of this paper to detail the results of the LRF performance during the individual smokes, but some generalizations can be made. Except for extremely dense concentrations, phosphorus smokes had little effect on the laser rangefinder. The particulate smokes, such as graphite and aluminum tended to inhibit ranging through the smoke, but the cloud itself could be ranged to. The oil based smokes were mixed results. Fog oil and JP8 seemed to be less of a problem, while diesel was a good obscurant to the 1.54-micron laser. It was also found that for the natural occurring obscurants such as sand, dust, and precipitation in any form could defeat the laser ranging to the target if the target could not be seen visually.

During Smoke Week XI, in Valcartier, Canada, several opportunities for testing the effectiveness of the laser rangefinder in the snow were achieved. It was found that under moderately heavy snow (visibility of 5-km) that the SIRE could range consistently. It was also found that the SIRE could range to the target whenever it could be seen visually. An example range performance in snow is shown Figure 4. The figure plots 1.54-micron LIDAR signal as a function of time/range. The LIDAR device is a separate instrument also used at the Smoke Week trials to provide quantitative data on obscurant density. The large peak starting at zero and decaying with time is the backscatter from a moderately heavy snowfall. Structure in the signal is indicative of density/reflectivity variations in the snow with position. The sharp "spike" near 15-microseconds is the target reflection. The corresponding SIRE range data (2325-meters) is shown in the upper portion of the figure. Even though the atmospheric conditions were very poor, and target contrast was low, the SIRE accurately determines the target range and disregards the background "optical noise".

TABLE 2
OVERVIEW OF SMOKE/OBSCURANT TEST RESULTS

Obscurant	Range to Cloud	Range through Cloud
Snow	Yes	Yes
PWP ZUNI	Yes	Yes
WP Wedge	Yes	Yes
LOFs, SRBOS, WRM LOF, etc	Yes	Yes*
JP 8	Yes	Yes
Fog Oil	Yes	Yes
Kaolin	Yes	Yes
Kaolin/Fog Oil	Yes	Yes
Kaolin/Canadian	Yes	No
Diesel	Yes	No*
Diesel/Canadian	Yes	No
Graphite	Yes	No*
Graphite/Fog Oil	Yes	No
Aluminum	Yes	No*
Brass	Yes	Yes*
Brass/Fog Oil	Yes	No
Brass/Kaolin	Yes	Yes

* Performance was more dependent upon smoke density than usual, with very dense smoke typically required to prevent ranging.

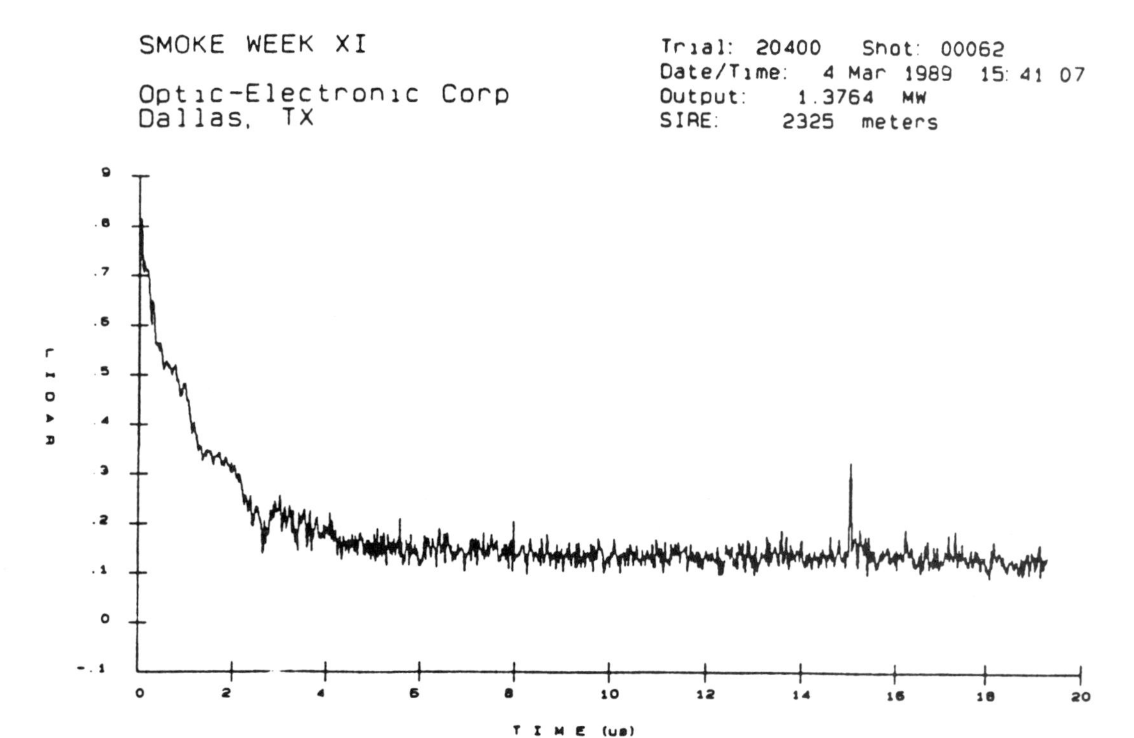

Figure 4 - Even in the presence of most natural and artificial obscurants (snow shown here), the SIRE can provide accurate range to any target that can be located using the 7-power sighting system.

6. CONCLUSIONS

The SIRE unit has been tested and proven as an effective tool for increasing the accuracy and effectiveness of armored vehicles in the battlefield. Improvements in time-to-fire, first-round hit probability, and overall gunner effectiveness of 300% are possible. The maximum range and range accuracy characteristics are more than adequate for all current and near-term armor missions. With the ability to penetrate both natural and artificial obscurants, the Er:glass wavelength is a principle benefit for real-world scenarios, and allows consistent performance even under adverse weather conditions.

Its excellent performance and design features notwithstanding, the key advantage of the SIRE is its complete eyesafety under all conditions. Close combat conditions are easily supported, even when hostile, friendly, and civilian factors are closely intermingled. Covert actions can be safely completed without concern for detection by conventional warning devices or sensor systems. Finally, a single fully-functional device can be used equally well for training and combat - truly supporting the philosophy of **"Train like we fight and fight like we train"**.

7. REFERENCES

1. "Occupational and Environmental Health - Control of Hazards to Health from Laser Radiation", Technical Bulletin - MED 524, Headquarters, Dept of the Army, 20 June 1985. (TB MED 524)

2. M.A. Woodall, J.R. Minch, J. Nunez, and H.S. Keeter, "Performance of Eyesafe Erbium:Glass Laser Rangefinders Through Natural and Artificial Obscurants", Proc. Smoke/Obscurants Symposium XIII, 25-27 April 1989, Johns Hopkins University, Laurel, MD.

Mid-Infrared laser applications

John G. Daly

Schwartz Electro-Optics, Inc.
3404 N. OBT, Orlando, FL 32804

ABSTRACT

Revived interest in mid-IR lasers can be attributed to better materials, medical applications, and their eyesafety. Twenty years ago, these 2 to 3 micron lasers were limited to research labs because of the necessity of cryogenic cooling. Recent advances have made room temperature operation with manageable thresholds available for $Co:MgF_2$, Tm, Ho, and Er covering from 1.7 microns to 2.5 microns. The interest in medical applications is related to the high absorption for tissue at these wavelengths and the flexible delivery by low loss fiber optics. The eyesafety issue makes commercial uses more attractive. Other interesting applications where eyesafety is critical include: rangefinders, remote sensing, wind shear detection and lidar.

1. LASER MATERIALS

Early investigations of rare earth ion doped crystals included potential candidates for mid-IR lasing. Johnson, Geusic, and Van Uitert (1) reported in 1965 on Yttrium Aluminum Garnets (YAG) with the trivalent ions: Thulium (Tm), Holmium (Ho), Ytterbium (Yb), and Erbium (Er). They included co-doping studies and chromium (Cr) sensitization which was successful in Tm,Cr:YAG for room temperature operation but with a threshold of 640 Joules. By the 80's more interest in eyesafe laser rangefinders that fall between 1.4 and 2.1 microns helped re-kindle interest in improved crystal growth. Er:glass, Er:YLF, and Ho:YLF benefitted from these efforts.

Throughout the 80's extensive work with chromium (Cr) sensitization of Tm, Ho, and Er in YAG and YSGG (Yttrium Scandium Gallium Garnet) hosts proved successful in developing room temperature lasing at lower threshold. The Soviets have made significant contributions to this work (2).

Holmium lasers have become a popular candidate for the 2.1 micron region. A combination of Ho, Tm, and Cr in both YAG and YSGG hosts provides optimum performance. Efficient flashlamp pumping of the Cr ions results in a nonradiative transfer to the Tm ions and subsequent transfer to the Ho ions. The laser transition is between the $^5I_7$ and $^5I_8$ levels. This laser crystal is commonly referred to as CTH:YAG or CTH:YSGG.

Thulium lasers also employ chromium as an efficient sensitizer. Cr,Tm:YAG (CT:YAG) lases between 1.9 and 2.05 microns. The transition from $^3H_4$ to $^3H_6$ has higher gain near 2.01 which will dominate in a simple resonator. YLF (yttrium lithium fluoride) has also been used as a host for Tm. Much of the interest in Tm has centered on its high absorption at 785 nm which makes it an excellent candidate for diode pumping. Diode pumping eliminates the need for the chromium. Holmium also benefits from this absorption at 785 nm since a Tm, Ho:YAG crystal can be diode pumped for an output at 2.1 microns.

Erbium in YAG and YSGG offers longer wavelengths with 2.94 and 2.79 microns, respectively. Er:YAG with dopant concentration of 50 atomic % can be flashlamp pumped at room temperatures. For lower thresholds, Er,Cr:YSGG should be used. The lower thresholds of the YSGG host are attractive for lower energy, higher rep rates, and Q-switching.

Although, it is not a rare earth element, Cobalt in magnesium fluoride is a solid state crystal with many similar applications. Co:MgF_2 is tunable from 1.7 to 2.5 microns. The first room temperature operation was demonstrated by Welford and Moulton (4). The fluorescent lifetime of lasing transition from $^4T_2$ to $^4T_1$ is strongly temperature dependent. Practical uses of Co:MgF_2 employ TE (thermo-electric) cooling of the crystal to maintain temperatures between 15 and 20 C.

2. LASER PERFORMANCE

These studies were conducted at both Schwartz Electro-Optics facilities: the Research Division in Concord, MA and the Solid State Laser Division in Orlando, FL. SEO lasers were employed with either a normal single mess PFN (pulse forming network) or the SEO Variable Pulsewidth Power Supply. The laser cavity was a close coupled elliptical design with water as the coolant. Normal use employed a refrigerated cooling system which operated with the water at 20 C. The laser resonator length was typically 30 cm and utilized a flat/flat mirror geometry. Q-switching in the 2 to 3 micron region has been accomplished by three techniques, a mechanical or rotating mirror, an acousto-optics (AO) module, and an electro-optical device.

2.1 Holmium

Holmium in CTH:YAG has found the widest application of these crystals. High energies and respectable average powers can be obtained. However, the CTH composition results in thermal problems associated with the Cr absorption. Laser rod thermal lensing is 4 to 5 times that found in Nd:YAG. This presents problems for high average power inputs which not only threaten resonator stability but also increases the temperature of the crystal. And since the holmium transition terminates at the ground state level, laser performance is very temperature dependent. Figure 1 compares the input/output performance for CTH:YAG, CT:YAG, and Er:YAG.

FIG. 1. Ho, Tm, Er COMPARISON

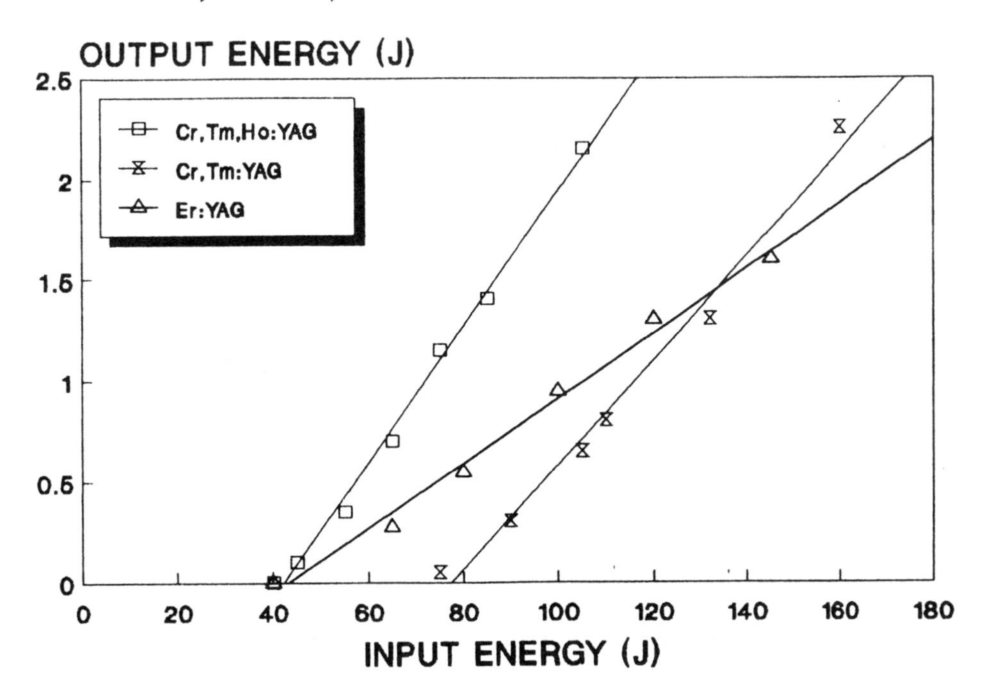

Most of the work with CTH:YAG has been long pulse (free running) with pulse durations from 100 to 600 microseconds. Per pulse energies to 5.0 Joules from a simple flat/flat resonator and repetition rates to 20 hertz for lower pulse energies produce average power in excess of 20 watts. Many of these tests utilized the SEO Variable Flashlamp Pulsewidth Power Supply which is well suited to optimization of flashlamp duration with performance. Figure 2 shows the laser pulsewidth control achievable with the variable pulsewidth power supply. This flashlamp driver delivers a square waveform to the flashlamp with duration adjustment via microprocessor control from 50 microseconds to 7. milliseconds. For CTH:YAG we have found that a 500 microsecond lamp duration provides the most efficient energy extraction.

Holmium can be Q-switched with a rotating mirror or an AO module. SEO has CTH:YAG laser systems available with up to 40 mJ per pulse at rep rates less than 10 hz and with 200 nsec durations. Electro-optical Q-switching has been accomplished but with limited success. Utilizing a brewster cut $LiNbO_3$ pockels cell we achieved 20 mJ in a 90 nsec pulse. However, performance suffered from pulse to pulse stability and secondary pulsing. These studies are continuing.

2.2 Thulium

In spite of lower gain and higher threshold than CTH:YAG, CT:YAG has potential applications attributable to its shorter wavelength (2.01 vs 2.1 microns) with a water (and tissue) absorption nearly two (2) times greater. Thresholds of nearly 80 Joules are twice that of comparable CTH:YAG material. But reliable room temperature performance can be obtained for input levels of slightly more than 100 Joules. We have demonstrated outputs exceeding 2.0 Joules at low repetition rates and 5 Watts of average power for 15 hertz systems.

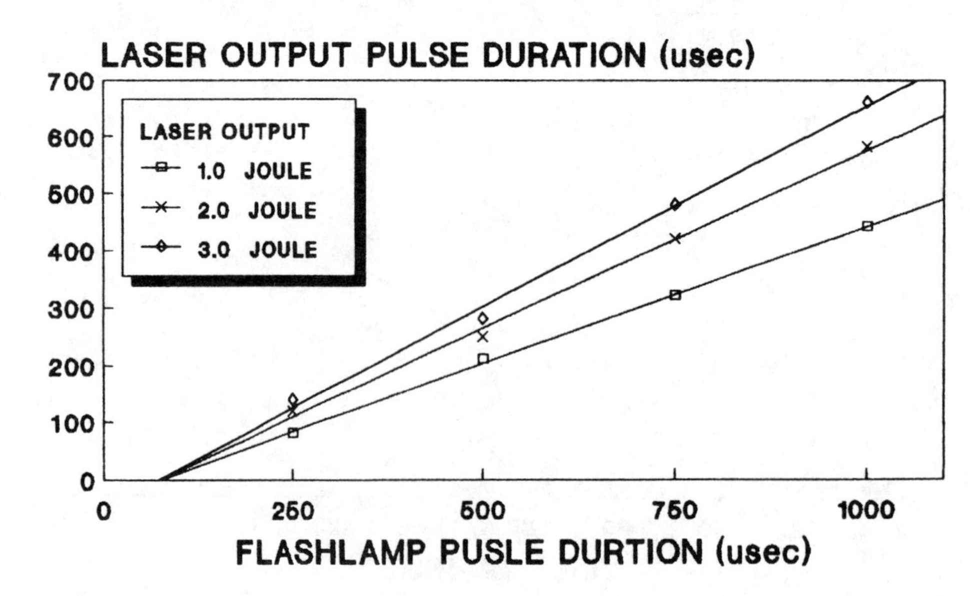

Q-switching studies are scheduled in the next few months. Parameterization of the flashlamp pulse duration has not been completed but preliminary indications are that 1,000 microseconds is optimum. We have also scheduled diode pumping studies of both thulium and holmium in YAG and YLF. Esterowitz (3) recently published an excellent review and detailed study of diode-pumped Ho, Tm, and Er. Simulated diode pumping studies are now underway at SEO with a titanium: sapphire laser tuned to 785 nm as the source.

2.3 Erbium

Er:YAG has a threshold of 40 Joules with a slope efficiency of 1.7% for a $\frac{1}{4}$ x 4 inch laser rod. The thermal lensing for Er:YAG is approximately twice that of Nd:YAG. The smaller 5 x 75 mm Er:YAG rod has shown a threshold of less than 20 Joules. To achieve lower thresholds, Er,Cr:YSGG should be used. A threshold of 2.5 Joules for a 5 x 75 mm rod makes this an ideal candidate for higher rep rate performance. We have shown an output of 80 mJ for an 11 Joule input at a 50 hertz rate.

A mechanical or rotating mirror Q-switch is easily implemented for the Er:YAG or Er:YSGG. Pulsewidths of approximately 200 nsec are available for pulse energies to 50 mJ for Er:YAG. Attempts to generate near TEMoo pulses from Er, Cr:YSGG lead to outputs of 20 mJ - 100 nsec pulsewidths using a rotating mirror at 400 rps.

We have recently Q-switched Er:Cr:YSGG with an electro-optic pockels cell. A $LiNbO_3$ brewster cut crystal provided more than 50 mJ per pulse in an 80 nanosecond duration for a 30 Joule input. Figure 3 summarizes the performance with normal mode and Q-switched data shown as well as the laser pulsewidth.

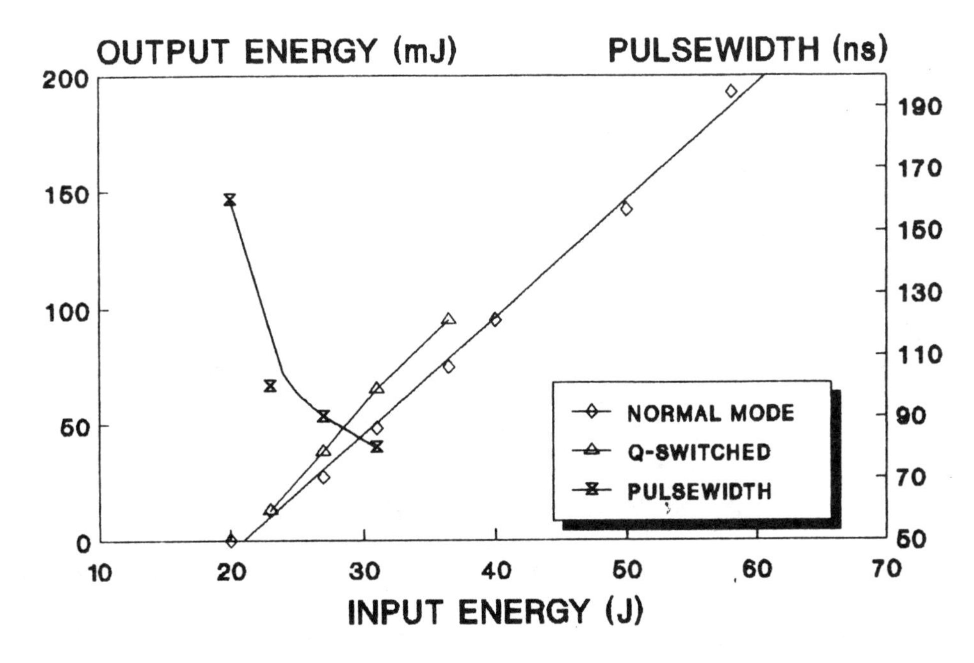

2.4 Cobalt

$Co:MgF_2$ is continuously tunable between 1.5 and 2.5 microns and has been demonstrated to operate both CW and pulsed. Early work by Johnson et al (5) at cryogenic temperatures used flashlamp pumping. Moulton (6) demonstrated liquid-nitrogen CW operation and TE cooled pulsed operation in 1985. Moulton used laser pumping with a 1.32 μm Nd:YAG source to achieve tuning from 1.75 to 2.5 μm at 10 Hz with 70 mJ per pulse output. As a result of this work and the room temperature work of Welford and Moulton (4), SEO developed a $Co:MgF_2$ product, the COBRA laser which delivers 100 mJ at 10 Hz with a 80 μsec pulse duration. As a laser pumped configuration, the COBRA laser easily attains TEMoo modes and can be scaled to higher energies and higher rep rates with the pump laser as the only limit. Figure 4 shows the broad tuning range of the COBRA laser. Each of the three curves represents a separate mirror set. SEO has also delivered CW $Co:MgF_2$ lasers which required a liquid-nitrogen dewar to maintain the crystal at low temperatures.

Q-switching of $Co:MgF_2$ has been achieved at more than 20 mJ per pulse. Improved coatings and slight resonator modifications are scheduled for more reliable operation at higher energies. The wide tuning range of this laser makes it an ideal candidate for remote sensing of many gases such as H_2O, CO_2, CO, CH_4, N_2O, HF, HI, and NO_2.

FIG. 4. COBRA 2000 TUNING CURVE
Co:MgF2

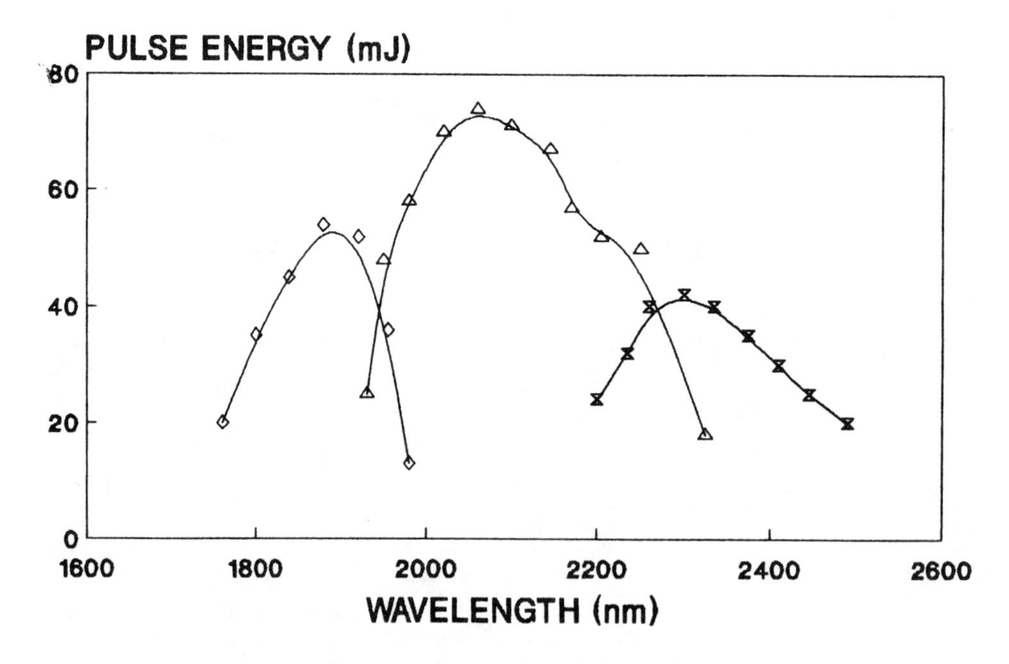

3. APPLICATIONS

As the mid-IR lasers become more efficient and are available in smaller packages, they are finding more uses. The medical field accounts for the largest volume at this time. Research is continuing in tissue interaction studies to consider the effects of wavelength, pulse duration and average powers. At this conference, they have dedicated an entire session, **Biomedical Optics,** to these studies.

The reader is referred to these proceedings for a current review of several mid-IR laser applications in medicine.

Laser rangefinders still represent an active and competitive market. The Nd:YAG rangefinders have dominated the military market. Recent trends to eyesafety have been lead by Er:glass and raman shifted Nd:YAG both operating at 1.54 μm. Early interest in holmium rangefinders may be resurrected if efficient electro-optic Q-switching is demonstrated. By the end of 1991, the future for holmium in the rangefinder market should be determined.

Remote sensing and LIDAR offer the greatest potential for the mid-IR laser crystals. At this meeting, Conference 1416, **Laser Radar VI**, has dedicated two and one half days to LIDAR applications which includes several Mid-IR laser discussions. The reader is encouraged to review these papers found in these proceedings. Solid state lasers capable of diode pumping present a rugged system configuration capable of OEM installations. Spaceborne sensing can also be accommodated by this laser package. Both the holmium and thulium lasers have been diode pumped and are leading candidates. The Er lasers have too much water absorption to function well as LIDAR sources.

The improved eyesafety over existing 1 micron systems is a strong advantage. However, the potential tunability over regions where differential molecular absorption can be utilized for sensing is a primary motivator. $Co:MgF_2$ is an ideal source for these uses. As laser performance improves, $Co:MgF_2$ will become a dominate laser in this field.

Therefore as the solid state laser community continues to improve these mid-IR laser systems, more applications will be found. The advantages of these wavelengths include: improved eyesafety, their solid state configuration, the high water (and tissue) absorption, their tunability, their adaptability to diode-pumping, and their scattering and molecular absorption characteristics.

4. ACKNOWLEDGEMENTS

These studies have been conducted at Schwartz Electro-Optics, Inc. Much of this work was internal R & D or was associated with studies funded by several SBIR programs. Contributions from the staff at the SEO Research Division in Concord, MA are recognized especially Peter Moulton, John Flint, David Rines, and Glen Rines. Ed Adamkiewicz, Madhu Acharekar, and Tom Kaffenberger of the SEO Solid State Division in Orlando, FL also made significant contributions.

5.0 REFERENCES

1. L.F. Johnson, J.E. Geusic, and L.G. Van Uitert, "Coherent Oscillation from Tm, Ho, Yb, and Er ions in Yttrium Aluminum Garnet," App. Phys. Lett., vol.7, no.5, Sept. 1965.
2. B.M. Antipenko, A.S. Glebov, T.I. Kiseleva, and V.A. Pis'mennyi, Sov. Tech. Phys. Lett. 11, 284 (1985).
3. L. Esterowitz, Opt. Eng. 29, 6. (1990).
4. D. Welford and P.F. Moulton, Opt. Lett. 13, 11 (1988).
5. L.F. Johnson, R.E. Dietz, and H.J. Guggenheim, Appl. Phys. Lett. 5, 21 (1964).
6. P.F. Moulton, IEEE J. Quantum Electron. QE-21 1582 (1985).

High repetition rate Q-switched Erbium glass lasers

Scott J. Hamlin, John D. Myers, Michael J. Myers
Kigre, Inc.
100 Marshland Road, Hilton Head, SC 29926

ABSTRACT

Many applications exist for eye safe lasers operating at high repetition rates. This paper will discuss the operation of Q-switched Er:glass lasers at high repetition rates with peak powers in the megawatt range.

1. INTRODUCTION

Laser emission from the $^4I_{13/2}$ - $^4I_{15/2}$ transition of Er^{3+} doped glass was noted as early as 1965[1], however, until recently there were few commercially available Er:glass lasers and virtually no Q-switched Er:glass lasers. During the last several years, a number of manufacturers have offered eye-safe hand-held rangefinders that utilize Er:glass Q-switched lasers.[2,3]

As the major producer of Er doped laser glass, Kigre has devoted a large portion of its internally funded research and development effort towards the study of Er:glass, Er:glass lasers, and Er:glass Q-switching technology. This paper will highlight some of the technology that evolved during the study of high repetition rate electro-optically Q-switched Er:glass lasers at Kigre.

2. YB, ER AND CR, YB, ER DOPED PHOSPHATE GLASSES

Figures 1 and 2 illustrate the absorption spectra of QE-7 Yb,Er:phosphate laser glass and QE-7S Cr,Yb,Er:phosphate laser glass. In both plots, QE-7S has the higher absorption. From Figure 1, it is evident that QE-7 absorbs very little throughout the visible portion of the spectrum (400 - 900 nm). The only strong absorption in QE-7 is between 900 nm and 1 µm, due mainly to the $^2F_{7/2}$ - $^2F_{5/2}$ transition of Yb^{3+} (Figure 3). Unfortunately, the absorption of Yb^{3+} does not match the emission of either Krypton or Xenon gas discharges very well, making flashlamp pumping quite inefficient[4]. From experimental measurements, it is estimated that only about 2 percent of the capacitor bank energy will result in useful pump energy for QE-7 when using a 1200 torr Xe flashlamp with a 3 mm bore and 45 mm arc length when pumped with a 15 Joule pulse for 1.5 ms.

Yb,Er:phosphate glass lasers are three level lasers when lasing at 1.535 µm, therefore, it is necessary that the population of the $^4I_{13/2}$ metastable energy level exceed the population of the $^4I_{15/2}$ ground state energy level. In order to calculate the zero loss condition (at the lasing wavelength) for a particular Er^{3+}

doping level, several assumptions may be made. Since the decay rates for the upper levels of Er^{3+} ($^4I_{11/2}$ and higher) are several orders of magnitude shorter than the decay rate of the $^4I_{13/2}$ metastable energy level[5], it is a reasonable assumption that all of the Er^{3+} ions are in either the $^4I_{13/2}$ metastable or the $^4I_{15/2}$ ground energy levels. The zero loss condition will be achieved when the metastable and ground state energy populations are equal, assuming there are no other losses in the glass. In QE-7 and QE-7S glasses this will occur when the population density of the metastable energy level is equal to half of the total Er^{3+} ion concentration (6.0×10^{18} ions/cm^3). In a common size Er:glass rod of 3 mm diameter by 50 mm length, the energy required to reach the zero loss condition is about 275 mJ.

In order to increase the flashlamp pumping efficiency, Kigre developed QE-7S Cr,Yb,Er:phosphate laser glass[6]. QE-7S is similar to QE-7, except that QE-7S contains Cr^{3+} as a sensitizer. Cr^{3+} absorbs flashlamp radiation in two broad bands centered at 450 and 640 nm, emitting in a broad band centered at 760 nm. This allows energy to be transferred from Cr^{3+} to the $^4I_{9/2}$ and $^4I_{11/2}$ states of Er^{3+} and the $^2F_{5/2}$ state of Yb^{3+}. This energy transfer makes QE-7S substantially more efficient for flashlamp pumping.

In addition to Nobel gas discharge pumping of Er:glass, several methods of laser excitation have been studied. Many authors have reported using Nd^{3+} doped lasers as optical pumps for Er:glass[5,7-8]. Anthon recently reported using a laser diode pumped Nd:YAG laser to pump a CW Er:glass laser[9]. Hutchinson also recently reported direct diode laser pumping of Er:glass[10]. Hutchinson used a GaAs/AlGaAs graded index, separate confinement heterostructure device to pump his Er:glass laser.

3. Q-SWITCHING OF ER:GLASS

Although the cross section of Er:phosphate glasses are considerably higher than Er:silicate glasses, they are still quite low in comparison to Nd:glasses. QE-7 has a cross section of about 8×10^{-21} cm^2. Although Er:glass has a low cross section, a three level lasing system, and difficulties in optical pumping, many researchers have been able to Q-switch Er:glass lasers.

The first Q-switched Er:glass lasers utilized mechanical devices such as rotating mirrors or porro prisms to change the Q of the resonator cavity. Currently, rotating prisms are still the most popular method of Q-switching an Er:glass laser. The MELIOS transmitter, which is the largest production quantity Er:glass laser, also utilizes a rotating prism as a Q-switch[11]. NEC recently introduced a hand-held eyesafe laser range finder that utilizes a frustrated total internal reflection (FTIR) device to modulate the cavities Q[3].

Kigre has been experimenting with electro-optical Q-switching of Er:glass lasers. Two basic resonator designs are used depending upon the repetition rate of the laser. If the repetition rate is low, a $LiNbO_3$ Pockels cell with a transverse field is used. The $LiNbO_3$ crystal is cut with Brewster faces, so that it also functions as a polarizer with an extinction ratio of approximately 10, which is sufficient due to the low gain of the Er:glass.

When operating glass lasers at repetition rates of more than one pulse per second, thermal induced birefringence becomes evident. In order to lase with a glass laser at higher repetition rates, a polarization insensitive Q-switch must be used. Researchers at Kigre have developed a proprietary electro-optic polarization insensitive Q-switch for high repetition rate lasers. This Q-switch will be presented at a future electro-optic conference.

4. KIGRE'S XE SERIES OF Q-SWITCHED ER:GLASS LASERS

Kigre has developed and markets several laser systems, operating at 1.535 µm, that utilize QE-7S Cr,Yb,Er:phosphate glass. The XE-100 laser utilizes the Brewster Q-switch described in Section 3. It is air cooled and capable of operating at ten pulses per minute, with about 15 mJ of output in a 40 ns pulse. The XE-100 laser rod is 3 mm diameter by 30 mm long.

The XE-200 and XE-300 laser systems utilize the polarization insensitive Q-switch, and operate up to two and four pulses per second respectively. Both systems actively cool the laser rod and flashlamp with liquid coolant. The pulse energy of both systems are about 15 mJ in less than a 50 ns pulse. Several XE-200 laser systems are currently being used in atmospheric studies and a XE-300 system is in use as an experimental Lidar transmitter.

5. CURRENT AND PLANNED WORK

Currently, Kigre is designing the next XE series laser system, which will be designated XE-400. In this unit, it is planned to use a simmered flashlamp and Kigre's newly developed variable pulse width flashlamp driver. This flashlamp driver delivers square wave pulses, continuously adjustable from 100 µs to 10 ms, allowing the flashlamps color temperature to be decreased for more efficient pumping. The XE-400 system's goal is five pulses per second with over a half megawatt of peak power.

The current limitation in peak power from Q-switched Er:glass lasers is the damage threshold of the $LiNbO_3$ Q-switch, the polarizer, and the antireflection coatings on the Q-switch and the polarizer. Researchers at Kigre plan to use the XE-400 as a test bed for testing more damage resistant optics, coatings, and resonator designs.

In addition to these resonators, Q-switches, and power supply developments, researchers at Kigre have been experimenting with the Er:glass itself. The first run of clad QE-7S laser rods was recently completed and is currently receiving promising results from tests in a XE-100 laser system. The cladding and core dimensions are chosen such that a tangential ray incident upon the outer diameter of the cladding will be refracted into a tangential ray incident upon the diameter of the core. This cladding makes the core appear larger in diameter, thus increasing the pump chamber's coupling efficiency.

New Er:glass composition experiments are also being conducted at Kigre in which both the base composition and the dopants are being varied. One of the main motives in these experiments is the development of Er:glasses optimized for laser and diode laser pumping.

6. CONCLUSION

Recently, there has been a high demand for Er:glass and Er:glass laser systems operating at the eyesafe wavelength of 1.535μm for the purposes of lidar, traffic enforcement, helicopter wire avoidance, wind shear measurements, communications, and other areas where human contact with the laser radiation is possible. Sales of QE-7N and QE-7S Er:glasses are now at an all time high, and it is believed that as the technology of Er:glass lasers become more mature, the demand will increase further.

7. ACKNOWLEDGMENTS

The research work at Kigre, referred to in this paper, was performed under various internally funded research and development programs. The authors wish to thank Valentin Gapontsev and Jiang Yasi for many informative conversations. In addition, the authors would also like to thank G. Rickel, R. Rickel, D. Rhonehouse, J. Harwick, B. Scott, and T. Rexrode for technical assistance.

8. REFERENCES

1. E. Snitzer, R. Woodcock - Appl. Phys. Lett., 1965, v. 6, p. 45.

2. R. Renairi, A. Johnson, "MELIOS: Status report of the U.S. Army's eyesafe laser rangefinder program," Laser Safety, Eyesafe Laser Systems, and Laser Eye Protection, Galoff, Sliney, vol. 1207, pp. 112-123, SPIE, Bellingham, 1990.

3. K. Asaba et al, "Development of 1.54μm near-infrared Q-switched laser," Laser Safety, Eyesafe Laser Systems, and Laser Eye Protection, Galoff, Sliney, vol. 1207, pp. 164-171, SPIE, Bellingham, 1990.

4. L. Nobel et al, "ILC Technology Tech. Rpt. ECOM-0239-F," October, 1973.

5. M. Zhabotinskii et al, Laser Phosphate Glasses, chap. 6, Nauka Publishing House, Moscow, 1980.

6. J. Myers, "U.S. Patent # 4,770,811," assigned to Kigre, Inc., Hilton Head, SC, 1989.

7. S. Hamlin, "Neodymium laser pumped Ytterbium activated Erbium doped phosphate glass laser," Digest of Topical Meeting on Advanced Solid State Lasers, pp 222-223, Optical Society of America, Washington, 1990.

8. V. Gapontsev et al, "Erbium glass lasers and their applications," Optics and Laser technology, **14**, pp. 189-196, 1982.

9. D. Anthon et al, "Erbium and Holmium Lasers Pumped with Nd:YAG," Optcon, 1990.

10. J. Hutchinson et al, "Diode pumped eyesafe Erbium glass laser," CLEO Postdeadline Papers, pp 638-639, Optical Society of America, Washington, 1990.

11. R. Renairi et al, "MELIOS - Status report of the U.S. Army's eyesafe laser rangefinder program," Laser Safety, Eyesafe Laser Systems, and Laser Eye Protection, P. Galoff and D. Sliney, 1207, pp 112-123, SPIE, Bellingham, 1990.

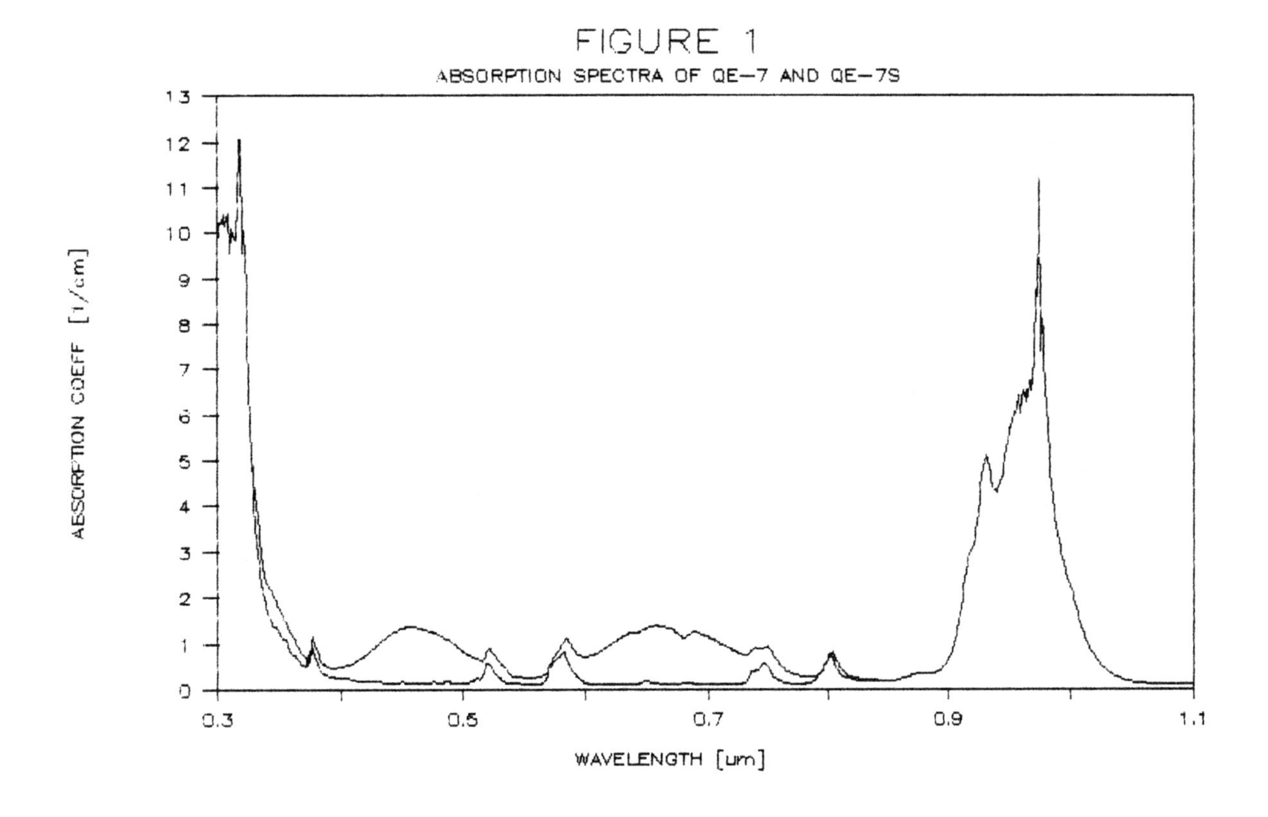
FIGURE 1
ABSORPTION SPECTRA OF QE-7 AND QE-7S
ABSORPTION COEFF [1/cm]
WAVELENGTH [um]
13
12
11
10
9
8
7
6
5
4
3
2
1
0
0.3
0.5
0.7
0.9
1.1

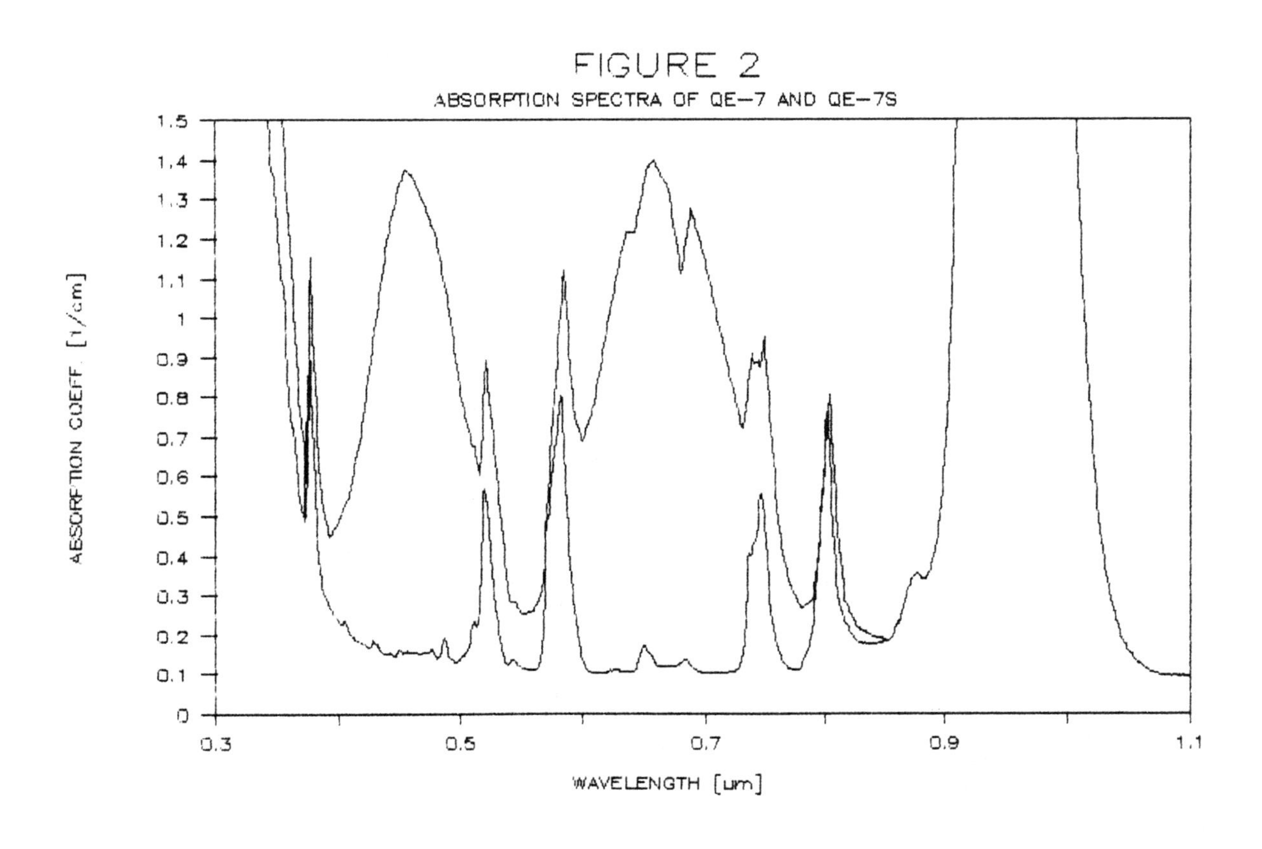
FIGURE 2
ABSORPTION SPECTRA OF QE-7 AND QE-7S
ABSORPTION COEFF. [1/cm]
WAVELENGTH [um]
1.5
1.4
1.3
1.2
1.1
1
0.9
0.8
0.7
0.6
0.5
0.4
0.3
0.2
0.1
0
0.3
0.5
0.7
0.9
1.1

FIGURE 3

Yb^{3+}, Er^{3+}:Phosphate Glass System

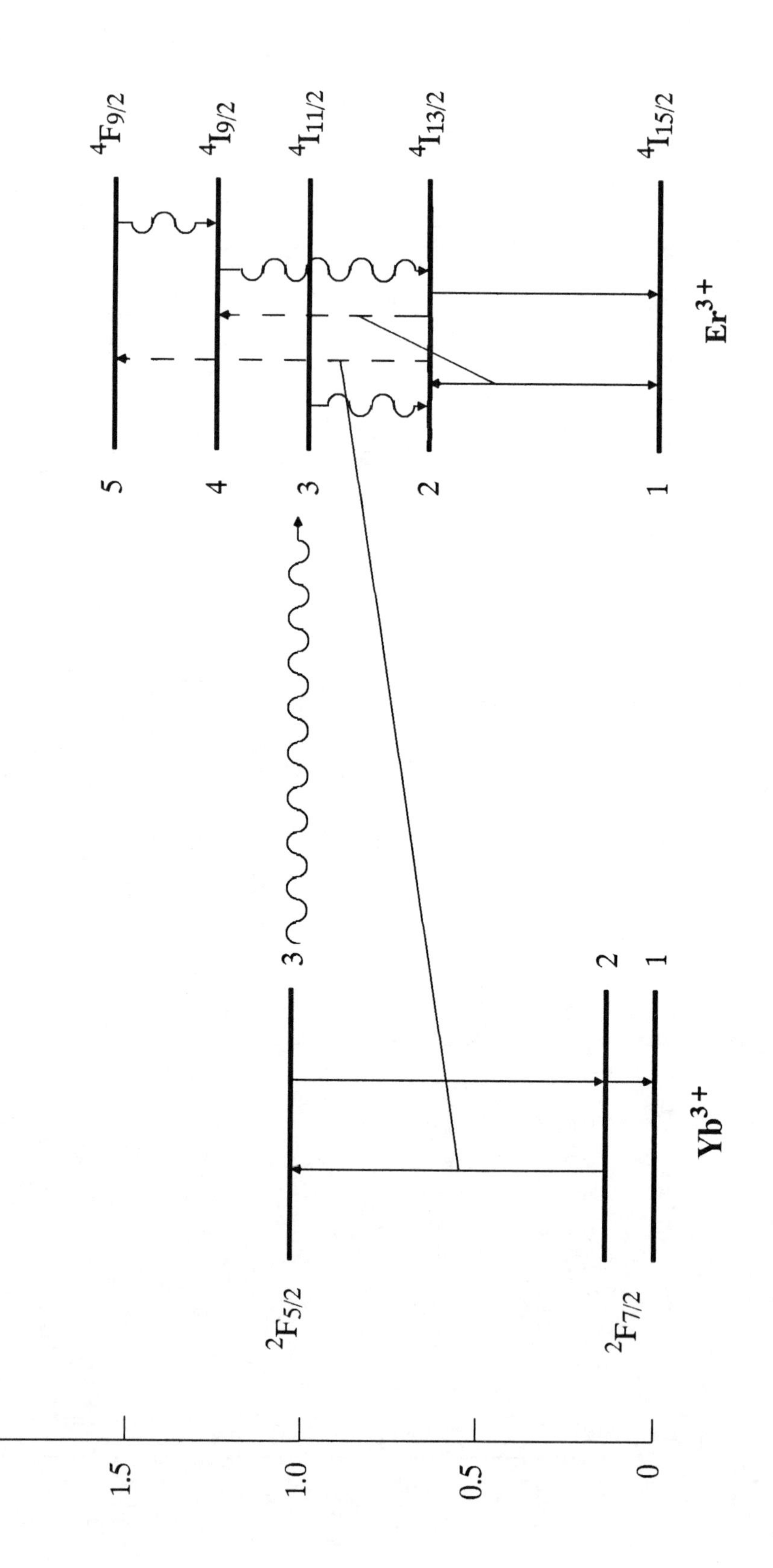

Eyesafe laser application in military and law enforcement training

Michael L. Mosbrooker

Loral Electro-Optical Systems

ABSTRACT

Training is a process of imparting a particular set of skills to a target group either by having them perform an actual task until proficiency is gained or by performing a similar task until confidence of proficiency is attained. Doing an actual task may be preferred but many factors may dictate that this objective is not feasible. The armed services and civilian law enforcement groups must train to use their weapons but often weapon characteristics, expense and the availability of appropriate facilities dictate that some sort of simulation be employed. Eyesafe lasers are playing a major role in this sort of simulation. Present uses include their employment as replacements for non-eyesafe lasers in determining the distance to a target, designating a target for laser energy seeking munitions and to signal the arrival of a munition at a target in a benign manner compared to what the replicated munition would do were it used instead.

1. INTRODUCTION

This paper describes the use of eyesafe lasers for military and paramilitary training. Paramilitary, as used here, generally refers to law enforcement agencies but also can apply to various physical security forces associated with both private and government agencies throughout the United States. Lasers described herein are not all Class I lasers and therefore not completely eyesafe under any conditions. Most of these lasers are Class IIIb. Lasers are considered suitable for training and are referred to as "eyesafe" when the prescribed precautions are transparent to and do not detract from the training in any manner. When the government evaluates the laser product of any contractor, a report is generated that will indicate the classification of the device in terms of eye safety and will narratively describe precautions that are to be taken and probable risks if any. For example, if a class III laser restricts direct viewing at a distance less than two meters and the physical configuration makes it impossible for a person to get to within two meters, the device is considered safe even though it is not technically Class I.

Simulation as a means of accomplishing almost any kind of training has been growing in popularity as technology advances make this form of education affordable and realistic. Simulation appropriately done will reduce training cost and increase training effectiveness in many situations and in fact provides the only alternative to training in the use of some of the expensive weapon systems we have today. Many weapon systems, particularly some of the missiles, are very good but are also very expensive to employ, and you simply can't afford to use enough of this expensive ammunition to train proficiently and sustain the training of the frequently changing crews. Another driver toward simulation is that the inherent danger associated with the use of many weapons systems requires that very elaborate and expensive facilities be employed if they are to be used in peacetime training. Finally, an element missing from the training of many military and paramilitary forces in the recent past is the ability to employ and measure the effectiveness of one force against another - a scenario ideally suited to today's simulation technology.

The training environment where we are using inexpensive eye safe lasers involves education in the employment of modern weapon systems of various types. This training is useful for both military and paramilitary forces and may be directed at maneuver or force employment type training or precision aiming systems designed to support individual or crew marksmanship under a variety of realistic conditions. Maneuver training embodies concepts associated with the tactical employment of a group of people against either another force or a dynamic scenario of various types, such as automated pop up targets.

Until the 1980's, the primary mode of maneuver training, particularly for the military with its powerful weapons, was what I refer to as the "bang you're dead, no you missed me" theory with very heavy use of controller personnel to attempt to sort things out. The sorting out was usually accomplished by noting which side had the preponderance of people and equipment in the engagement without regard to anyone's proficiency with the equipment. In fact in a typical army exercise, you might have one division as a maneuver force with another division or at least all the senior personnel from that division serving as controllers and directors of the exercise in order to assess casualties and maneuver effectiveness.

In order to use simulation as a viable training vehicle, it must be faithful to the performance of the system being simulated to the greatest extent possible. LA frequent limitation is cost where the question asked sooner or later is whether a 10% improvement in fidelity is worth a 90% increase in cost. The person developing the simulation of something like a weapon system has to walk a very fine line. The "goodness" of the simulation will be perceived differently by different parties. Invariably the weapon being simulated will be seen from the target's point of view as much more effective than the real system would be and the people using the weapon will think that the real system would actually be much more effective that the simulation. In fact this is what the Contractor needs to hear if he is going to survive - he must walk the middle ground.

In designing any training system, laser based or not, a critical goal is the avoidance of negative training. That is, the trainee must not be required to operate knobs, switches, lights, displays, etc. that differ from the actual machinery to be operated. Where surrogate switches are used, they must be in the same physical configuration and location as the real thing. Goals cannot always be 100% met within desired cost structures but the closer we get, the more effective and realistic the training. Thus, in military weapon systems, we use our eyesafe lasers to replicate as faithfully as possible the damage mechanisms but the maintenance and sustenance of the weapon platform and/or its crew must actually be performed during the training period just as would be required under actual employment conditions.

An important aspect of training is to be able to review the activity and thus to learn from errors and reinforce good behavior. Initially, the laser training systems were used one shooter vs. one target. Any results of the encounter would be apparent to these two parties but not necessarily anyone else. In order to expand the view of activities to all players, instrumented ranges are being developed. The eyesafe laser remains one of the primary mechanisms which determines outcome for a given scenario. However, on the instrumented range, it is possible to achieve real time feedback of exactly what is occurring as the maneuver scenario is being played out. Thus feedback will usually be presented on a display of some sort. Of greater importance is the ability to record and play back the complete portrayal of events to include variable speed and zoom. Thus all parties to the exercise can determine what techniques may be improved at any level, including the overall maneuver strategy to individual operator proficiency.

Instrumented ranges also include the ability to take a through-the-sight (aiming device) video picture of exactly what the various "gunners" are looking at, at the time they "fire" their weapon. In this way, the resulting video provides not only useful feedback for training but also precludes a lot of excuses and pointless arguing. The resulting video may be synchronized with the graphic replay so that both may be viewed simultaneously during and after action review and critique.

Event recording is also included with instrumented ranges so that after the fact determination can be made of who employed his weapons, with what effectiveness, and what techniques were used such as modes of operation and the switchology used to control the mode chosen. At the end of the exercise or other time of the exercise controller's choosing, this event material from each player unit may be dumped to computers where it is compiled in a format of the exercise controller's choice. Note that this mechanism will also serve to point out any weaknesses in the simulation system itself - thus the training system contractors police themselves.

2. TECHNICAL CONSIDERATIONS FOR TRAINING SYSTEM LASERS

The role of eyesafe lasers as part of the type of weapon training device falls into two basic categories. The first of these is to use the eyesafe laser to replace non-eyesafe lasers associated with weapon systems to be simulated. The second is to use the eyesafe laser as a surrogate bullet or missile. In the first category, several systems that are simulated use Neodymium YAG (Nd YAG) lasers as range finders and/or designators. Laser designators are typically modulated pulsed lasers used to "paint" a target such that the incoming weapon will seek the reflected information and energy. The rangefinder may be used to determine at what time to turn on the designator so the munition can locate its desired point of impact.

The Neodymium YAG lasers thus used are very powerful and are certainly not eyesafe. As a result there are very few places where the operator is even allowed to turn this equipment on much less use it to train in any kind of maneuver scenario. By replacing this laser with one or more eyesafe lasers, training can be conducted with manageable safety constraints. In a typical situation where a Nd YAG laser is used for range finding and designation, the range finding function is simulated by an Erbium Glass laser and the designator function by a Gallium Arsenide laser. The two lasers used to replace one poses some very interesting packaging problems. In the second role for eyesafe lasers, the laser is used in place of a bullet or other munition. The laser strikes the target instead of a bullet and the effect is assessed by means which will be discussed later.

The range performance required of an eyesafe laser must be equivalent to the effective range of the weapon simulated. Generally the weapons simulated are restricted to anti-tank missiles and firearms where one or more system operators are expected to directly view the target before firing. It is interesting to note that when eyesafe lasers were initially used in simulations, required ranges have grown from less than four kilometers in the early 1980's to approximately eight kilometers now - with no change in eyesafety criteria. This situation may become even more challenging as new weapons systems achieve longer and longer ranges.

Eyesafe range finders referred to are typically made using Erbium Glass technology as developed by Optic Electronic Corporation. Such range finders are employed at ranges out to ten kilometers with accuracies on the order of plus or minus five meters. Another technology used to provide eyesafe range finding is Raman shifted Nd YAG, which has been extensively tested but to date has not proven superior to the Erbium Glass laser.

The lasers that will serve as the focus of this paper will be the Gallium Arsenide lasers used as designators and to simulate the impact of a munition. In any of the simulations being done to date, the GaAs laser actually serves in the latter role as there is no true designation function to be performed. The foundation of the GaAs laser is the GaAs diode. These are used in three and six mil sizes, are inexpensive, efficient but nonetheless do offer some problems. Some of the drawbacks include limited obscurant penetration, poor temperature compensation and lot to lot performance not being particularly consistent.

Having said that, it should be pointed out that none of the systems being simulated perform all that well under obscurant conditions. The poor temperature compensation is being rapidly overcome by new developments in junction diodes, and some of the bulky and somewhat grandiose temperature compensation schemes of the few years ago are very quickly being eliminated. The lot to lot efficiency could be a problem, but in fact the efficiency tends to get better with time and not worse. Thus, all that needs to be done is to restrict output to that originally approved by the government agency that looks at laser safety of such devices.

There are several sources of power that are candidates in various combinations to operate the laser based simulation systems. These include piezoelectric crystals, dry cell batteries and platform electrical systems. As the system is generally expected to operate (exhibit vulnerability characteristics) with the host platform shut down, dry cells are normally used. In pistols, a stack of three hearing aid batteries operate the laser with a piezoelectric crystal to trigger the laser. Vehicle mounted systems typically use dry cells for the laser and its controlling electronics with platform power being used for high current devices like a signaling strobe light.

In order to effectively employ an eyesafe laser in a training mode, it must be capable of carrying information to the target. This is accomplished by modulation of the laser beam. Pulse code modulation is used at a frequency of 3000-6000 Khz with an eleven bit word used to provide the desired information. Each valid eleven bit word contains exactly six "ones".

Where the weapon being simulated is fired electrically, the laser activates on a trigger pull signal from the weapon. The modulation may not be linear (words sent per unit time) at this point in order to correspond to effect desired - more on this later.

Small arms other than pistols, are activated by blank ammunition designed for the weapon. When the blank ammunition is fired, a microphone senses the acoustic perturbation which in turn causes the laser to activate. Pistols are also activated by blanks but here the blank activates the piezoelectric crystal, which then provides sufficient power to trigger the laser system.

The pulse code modulated laser itself is only one end of the scheme. There must be some sort of detection means on the other end if the effect of the laser is to be measured. Typical of systems in use today are arrays of simple silicon detectors which have sufficient sensitivity to match up the desired weapon characteristics and target vulnerability. The silicon detectors are placed geometrically with spacing sufficient to control scintillation effects but yet provide the degree of sensitivity needed for a realistic simulation.

Another area of weapon training where eyesafe lasers are used is in precision gunnery or marksmanship. Simulation of precision generally adds complexity as the emphasis is to refine the system to give information on just where a target was struck and with what result. Precision simulation is applicable to both small arms and to large caliber weapon systems.

For small arms, a finely focused beam is used, and the target will have an array of independent detectors. Detector spacing is such that any beam that strikes the target silhouette will also impinge on one or more detectors. This is in comparison to non precision systems where the detector signals will be added to produce a single cumulative result for the entire detector array. Each of the target players will have an event memory and will automatically record any shots fired or hits received as well as the resulting effects assessed.

Various types of sensors are used for sensing trigger pull on small arms. Sensors described previously often do not work in a precision aiming environment. The microphone won't work because of requirements for precision systems that stipulate that more than one weapon be in play and one weapon firing will not set off another at a distance of perhaps half a meter. Therefore, various alternative sensors are used in order to determine whether or not the weapon has been fired. Even blank round firing provides sufficient changes in acceleration to allow measurement by suitable detectors.

Some pistols, notably automatics, bring about additional problems in terms of simulation initiation. The blank which has sufficient energy to operate the automatic feed mechanism causes recoil, which makes it very difficult to transmit the quantity of information required at the frequencies used without the aim changing. There are various schemes in the works right now to compensate for this anomaly. One is to speed up or accelerate the frequency of modulation. However, the system is then no longer compatible with older systems with have slower modulation frequency.

Tanks systems, as you might expect, are somewhat more complex. Two areas to be simulated on precision tank systems which are neglected in the force-on-force of maneuver systems are simulation of the non-flat trajectory and also simulation of the required lead angle that someone must employ when firing at a crossing target. Precision systems also employ refined damage criteria for the target.

Non precision systems typically will determine whether or not a given event has resulted in a hit on target and if so, a probability table is consulted which to determine what a hit on this particular type of target might accomplish. The probability of damage is a certain number. A Monte Carlo run will be made and will determine whether or not that damage actually occurred. On a precision system a much more refined resolution is sought wherein the target will be divided into some sort of matrix with each element having its own probability of damage. Event memory will function as discussed previously.

A typical precision tank system will start the simulation at the time of trigger pull. At the time of firing, the pointing direction in space of the gun barrel will be sensed with gyros and accelerometers. The point in space where the fired shell would be is then continuously calculated and updated. Compensation is made for variability in muzzle velocity. As the position of round is calculated, a tracer trajectory will be injected and appear in the gunner's sight showing where the round is actually flying at any moment. The laser itself is directed to the point where the actual round would have been at any point in time.

Target vehicles will have combination detector and reflector units mounted on them. When the firing vehicle senses a reflection from any target, the firing platform will determine the point of impact on a target based on the frequency shift of the reflected coded beam as it returns from the target's retroreflectors. When the laser intersects any target, a simulated burst will be injected into the sight to depict round detonation. The target platform will be notified by laser modulation of the type of round and its point of impact. The target platform perform's its own damage assessment. The target will then react appropriately with smoke or a strobe light or other means to signal that it has been damaged.

3. LASER TRANSMITTER PHYSICAL CONFIGURATION

The physical configuration of various eyesafe laser transmitter packages differ from system to system. For pistols, we will have a cylindrical package approximately two inches of length with an outside diameter which approximates the bore diameter of the pistol. This unit is inserted in the forward portion of the pistol bore. The assemble contains the piezoelectric crystal, battery stack, modulating electronics, laser diode and lens. Each pistol designated for use with this type of transmitter is modified so that live ammunition may not be fired in the pistol once it is set up for training.

Small arms which include rifles and machine guns up to .50 caliber use a separate transmitter mounted co-axially with the bore of the weapon. This is a light weight device powered by a nine volt dry cell battery which provides sufficient power to lase to the effective ranges of these types of weapons.

On a large bore directly aimed cannon, such as the main gun on a tank, the laser transmitter is normally located within the bore of the cannon, near the breech end. This type of transmitter contains only the desired number of lasers and optics with the power and electronics located in separate interconnected consoles. Thus, the laser(s) will be coaxial with the cannon bore and will follow the pointing of the gun bore in non-precision systems. Within the fire control system, super elevation and lead angle must be disabled as the gun must be pointed directly at the objective to function properly.

The other type of weapon currently simulated is the missile. The transmitter configuration is similar to that used for the large cannon but the physical location of the transmitter can usually be in any convenient location on the launch platform, and in some cases is located with a completely different portion of the system than that which actually launches the missile. The only constraint is that the laser that will replicate the missile must have line of sight to the target at the expected time of arrival and for a period of time prior to arrival that will simulate the tracking window, usually about 10 seconds.

There are a number of drivers to be considered in selecting the appropriate physical configuration for the various types of weapon systems. Some of these include training realism - where a twenty pound package for a small handgun would not be acceptable. Production and life cycle cost are always major factors in any design. Interface requirements in terms of connections to instrumented ranges, event recorders and the like may constrain the location of supporting electronics. Power requirements vary due to the required transmission range of the laser for a particular weapon, the amount of support electronics used for the system being simulated and the required duty cycle.

4. DESIGN CONSIDERATIONS FOR A LASER BASED WEAPON TRAINING SYSTEM

In considering the design of a laser based weapon simulation, two parameters of particular interest are compatibility of all players and the characteristics of each player that the simulation system is to replicate. Each weapon type must realistically address each target type, and each target type must respond appropriately to each weapon type. Some optimization is required but in general, a specified quantifiable result is expected from each possible combination of engagement. For players as targets, the vulnerability of each type target as a function of a hit by each type of weapon must be known. For each weapon type, the accuracy and range capability of each type must be specified. All possible modes of operation to be simulated must be understood to include the effective range and rate of fire and, the type of weapon involved such as gun or a seeking, beam rider or fire and forget missile and whether the gunner must continue to sight on the target until the missile arrives. Finally, the integration of players must be such that when a new weapon is placed in inventory, it can be simulated without changing any of the existing player equipment.

Before arriving at a final configuration, a review of the constraints that we are to be faced with is appropriate. Many of these can and must be made to work for us in accomplishing the simulation. The first of these is the atmosphere. The atmosphere will cause considerable and variable attenuation as we attempt to determine the laser parameters to produce the desired effective range. Another is eyesafety. It might be possible with a large amount of laser power to do whatever we need to accomplish to do the simulation but the intent and one of the key goals of such a simulation system is that it be eyesafe, or at least eyesafe within a reasonable bound so that the system can be used without unusual precaution or negative training.

The operating environment the equipment will experience is an important consideration. Not in any particular order, there is the mechanical shock caused by the natural shock and vibration that a tank or plane carrying the simulation equipment will see throughout its life. The equipment will also see temperature extremes which range from an ambient minus 25 to + 62°C with the high end going to over 80°C because of the proximity of some of the transmitters to the barrel of the blank firing automatic weapons such as on several rifles and machine guns. And finally, an environment that can not be forgotten is the user. Some of the users who will be mounting, dismounting and operating this equipment can be something less than careful and in fact are capable of breaking just about anything that man is capable of building. The customers expect that our products will stand up to all of these environments, and to be successful, these environments must be considered throughout the design process.

Another design constraint is life cycle cost or the total cost of buying, operating, maintaining and ultimately disposing of a piece of equipment. In considering the life cycle cost, the initial procurement cost is usually minor except that the purchase price usually must be budgeted and paid up front. An example of a life cycle cost driver might be batteries, as anyone who has bought toys lately can attest. The cost of the batteries themselves may be significant, and if the system requires that trained technicians change the batteries, the labor cost will add up. Often, a small change in consumable spares on a system where many like items will be produced will make a significant dollar change in the life cycle costs for the system.

The customer for the eyesafe laser based weapon simulation system will often specify additional characteristics which then become design constraints. Some of these include the ability of a target to sense and report near misses as well as hits. The near miss aspect is to alert the target that it is under attack and if possible should take evasive action. False hits must be minimized and thus there are a specific number false hits per hour of operation that are allowed.

Power requirements are to be minimized, not only to save money, but to allow the system to operate without undue interruption. Most of the systems built require between 20 and 50 milliamperes when transmitting, depending on the laser configuration. In a standby mode, the current draw is approximately 2 milliamperes. An important concession granted by the customer was to specify a "standard" day against which atmospheric attenuation would be measured. This day was defined as one in which the visibility is 23 miles. However, this standard day gave no concessions as to the time of day or the temperature, both of which affect scintillation.

And finally the Customer provided the probability of a hit for each weapon-target combination, as a function of range and also the probability of damage given a hit on each possible pairing of target and attacking weapon possibilities. It was then the designer's job to appropriately implement the stipulated probabilities and prove that the system is faithful to the values specified.

Controllable variables available to the design team include transmitter power output which is variable within eyesafety constraints, laser beam divergence, the quantity and type of information used to modulate the laser and the threshold level of the laser detectors. The decoding scheme employed for the information transmitted, the type of detectors used on each target, the spacing and location of these detectors and finally the number of words that must be received in order to constitute a particular event are target parameters for each type of target that the designer has available to drive the desired results for the system.

5. DESIGN APPROACH

Successful integration of all of the constraints and specified parameters did not happen on the first pass. The simulation science based on eyesafe lasers is simply not that precise. Instead the development was evolutionary and started with prototypes based on a theoretical design which was then fine tuned and modified and ultimately proven satisfactory during tests conducted by a group of ultimate users.

GaAs diodes were selected as the heart of the laser system. Three mil diodes were selected for small arms and machine guns as being sufficient to provide the necessary beam profiles for these fairly short range weapons. Six mil diodes were chosen for the larger weapons such as tanks and missiles. Selection of the GaAs diode was based on several characteristics. They are compact, light, reliable, easily modulated, match the peak response of inexpensive silicon detectors and can operate in an eyesafe mode.

The largest lens employed on any eyesafe lasers to date is about 50 mm in diameter with a focal length of over 200 mm and F/2.5. The beam profile is elliptical, generally with the major axis oriented horizontally for most weapons. The size of the ellipse is generally about the size of the logical target. Actual beam size for a tank main gun laser is about 2 x 4 feet on the minor and major axes respectively. The beam size achieved is a function of the square root of the laser power divided by the detector's threshold setting. The beam waist occurs approximately 2.5 meters in front of the lens with the far field considered to start at 50 meters.

The word size selected for the simulation consists of eleven slots of which six are always "ones" or an eleven bit word with weight of six. There are no synchronization bits as the codes selected for the various weapons being simulated are unique. When words are connected head to tail, it is not possible, no matter where one's starts in the resulting string of bits, to receive any valid code other than that intended as long as bits are not added or subtracted. The standard weight of six assists in discarding invalid words where bits may have been lost or added. The pulse width used is between 100 and 200 nano seconds.

Decoding at the target of the information transmitted on the laser beam is accomplished using Boolean Union techniques in order to compensate for dropped (but not added) bits. Boolean Union decoding is accomplished by comparing the output of a shift register operating at the 16th harmonic or 48 Khz with the next string of eleven bits (one word). If either or both words have a one in a particular slot, the result send to the decoder is a one. The only situation Boolean union decoding will not correct is the unlikely case where there two dropped bits exactly one word length apart. Boolean Union decoding is particularly useful in the simulation scenario due to the fact that the binary communication channel is very unsymmetrical. Scintillation causes bits to drop out more frequently than false bits are added. Simply lowering the detector threshold would cause the false alarm rate to exceed specification. Thus, the Boolean Union decoder effectively combats high frequency scintillation while raising the false alarm rate only slightly.

The decoders used are always looking for a valid code to appear. The decoder is located just downstream from the Boolean Union and when a valid code is seen, the decoder counts this as one word received and looks for another word. Hits are assessed based on the detector seeing a required number of identical valid code words within a specific time frame, which is varied to achieve the desired probability of hit effects.

The number of words that constitute a hit or near-miss varies by system type. For small arms and their typical targets only one decoded word out of eight transmitted is required. For cannon or tank gun systems, two words are required while missiles may require decoding of up to 22 words out of perhaps 32 transmitted. With most of the missiles simulated, the requisite number of decoded words constitutes a hit and damage. Other systems must consult the probability tables to determine whether damage has occurred given a hit.

The detectors are inexpensive silicon and a sufficient number are used to reduce the scintillation effects to a manageable level. Detector spacing is always greater than the Fresnel zone size and thus, the signals received by each detector are considered independent. This spacing does cause close range misses as the laser beam can impinge between all detectors. Such an occurrence is referred to as pseudo miss. The combined effects of scintillation, multiple words/messages, detector location geometry, and electronic signal processing are used to achieve the desired vulnerability effect for each pairing of weapon and target.

The near miss effect is achieved by sending a larger number of near miss (vice hit) indicating words to the target and where appropriate, making a slight adjustment electronically to the power output of the laser to increase beam size. The near miss is thus structured to be a more likely event than a hit.

In order to make the laser based system fully compatible with the various instrumented range configurations, it is necessary to identify not only type weapons and targets but also to establish positive identification of the various players. This is accomplished by interlacing player identification information within the eleven bit word by making use of the 48 Khz shift register. Up to four total bits are added by using one of two positions in between the leading edges of existing weapon code bits. The total weight of the word is now ten rather than six. With this scheme, 5280 discrete player ID's are possible. One significance of this scheme in terms of system design is that the leading edge of some pulses may now be only 125 micro seconds apart vice 333 micro seconds without player ID.

6. EYE SAFETY

Laser eyesafety for this type of application must consider both aided and unaided viewing possibilities. The final position approved by the government customer is that only the small arms transmitter (SAT) of the GaAs laser systems considered for transmission of codes is Class I. Everything is either Class IIIA or IIIB. The Erbium Glass lasers used for range finding are considered Class I.

The naked eye restriction on the worst of the systems is 6 meters. That is manifested as a posted warning which is posted and explained telling users not to look directly at a laser at a distance of less than six meters or at all when using optical instruments. More specifically, the use of tank optics at ranges of less than 75 meters is not permitted unless protective filters are installed. Recommended filters are KG-3 where such filters do not already exist, as they do on the main battle tank and infantry fighting vehicle.

These restrictions are not considered to be unacceptably constraining by the users of this equipment. There are several mitigating factors. The six meter restriction is of particular interest only on missile systems, which are not particularly common in terms of the numbers of guns employed of all calibers. On individual weapons and machine guns that are operated using blanks, the safety restrictions on the blanks are more stringent than the six meter constraint. On the tank guns, the laser is mounted in the barrel at the breech end and the barrel is longer than six meters.

The government agency that evaluated these laser systems also believes that there is a safety factor of about twelve built into all of the safety criteria applicable to the simulation systems discussed here. All of the eyesafety calculations and evaluations were conducted in accordance with Army Technical Bulletin MED-279. All systems except the SAT were exempted from the Code of Federal Regulations, Title 21, Chapter I, Subchapter J, pursuant to Exemption No. 76 FL-01 DOD issued on 26 July 1976.

When the systems were approved for operational use, reports were generated that specify what power settings were used and thus formed the basis for the restrictions cited above. During production and in the development of simulators for newly introduced weapon systems, these levels must be adhered to unless there is a completely new evaluation by an appropriate agency. As the power level of the GaAs diodes continue to improve, it is necessary to restrict the level of input current to maintain an acceptable level of power output. In order to compensate for the variable and improving efficiency of the laser diode, every transmitter built has a "select-in-test" resistor which is set to match the output of each diode to the specification originally approved by government safety personnel.

7. CONCLUSIONS

o GaAs lasers, operating at eyesafe levels, have proven a tremendous success in providing the communications links essential for realistic weapon simulation systems.

o There is currently no other technology that can match the laser for short range point to point reliable communications at anything approaching the cost of the GaAs packages.

o As the effective range of the weapons being simulated continues to grow, the task of designing a laser based simulation will become increasingly more difficult.

BATTLEFIELD TRAINING IN IMPAIRED VISIBILITY

Rudolph R. Gammarino

Loral Electro-Optical Systems
600 E. Bonita
Pomona, CA 91767 USA

James W. Surhigh

U.S. Army, PM TRADE
12350 Research Parkway
Orlando, FL 32826-3224, USA

ABSTRACT

A laser training system entitled Shoot Through Obscuration MILES (STOM) is being developed to operate with Forward Looking InfraRed (FLIR) systems during battlefield exercises where visibility is impaired. The STOM system is capable of ranges in excess of 6 km and can penetrate battlefield obscurants such as fog-oil, smoke, dust, and rain.

1. INTRODUCTION

STOM is designed to complement the existing Multiple Integrated Laser Engagement System (MILES). MILES uses eyesafe GaAs semiconductor lasers emitting at 0.904 micrometers to simulate the various weapons being fired. Silicon detectors are placed on various targets to receive the incoming laser hits. However, MILES cannot successfully penetrate natural or manmade obscurants which limit visibility due to the short wavelength of the GaAs laser.

STOM employs an rf excited CO_2 laser which operates in the center of the FLIR's spectral window at 10.6 micrometers. The laser, which is sealed, non-cooled, and measures less than 6 inches in diameter and less than 2 inches high, delivers 30 W pulses. The pulse width, which can be as short as 50 microseconds, and the time between pulses are microprocessor controlled. In addition, the STOM system uses an extremely sensitive, room temperature laser receiver which consists of a pyroelectric detector coupled with a low noise hybridized preamplifier. This receiver, which has an active area of 1 cm^2, is capable of detecting the 10.6 micrometer STOM laser pulses with irradiance as low as 50 micro W/cm^2. The STOM system will permit a gunner using FLIR to engage otherwise obscured targets during training exercises.

The key to the success of this program was the development of a room temperature, relatively low cost 10.6 micrometer detector. A survey of available material led to the choice of a lithium tantalate pyroelectric detector. In addition to making a sensitive detector, a very low noise preamplifier was designed specifically for the pyroelectric detector's electrical characteristics.

A study was made to determine the effect of detector thickness on responsivity. Detectors with thicknesses of 10 microns, 25 microns, 100 microns, and 225 microns were manufactured and were then simultaneously coated with black gold to eliminate any variation in coating characteristics. The devices that were manufacturaed had an active area of 1 cm^2. Figure 1 shows the results of the investigation. The thinner the device, the higher the responsivity. However, the difficulty in manufacturing the devices is inversely proportional to the thickness. Extremely thin detectors are difficult to handle because they are so fragile. The practical limitation with present manufacturing techniques is estimated to be 25 micron thickness.

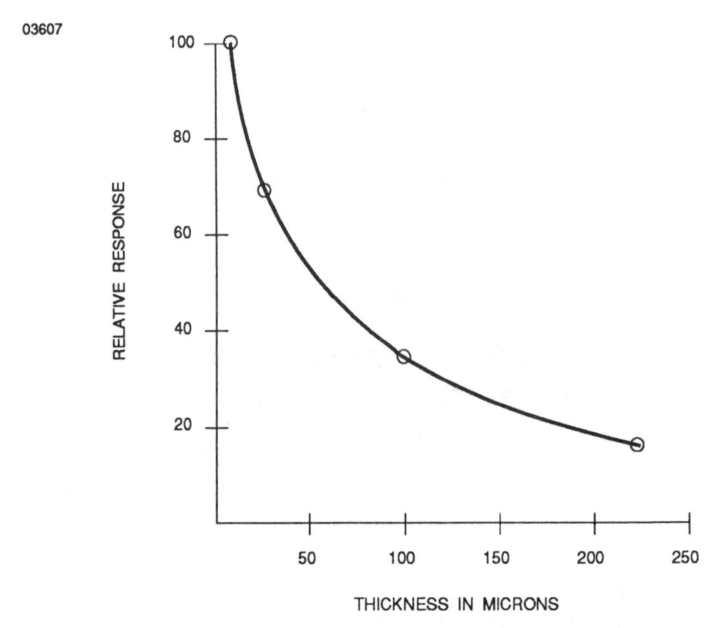

Fig. 1. Effects of Thickness on Detector Responsivity.

The combination of the pyroelectric detector and low noise preamplifier permits detecting 10 micrometer laser pulses at a level of 50 micro W/cm^2 with an extremely low false alarm rate. There is one technical problem which has yet to be addressed, that is the pyroelectric detector must be shock isolated from its mechanical package since all pyroelectric detectors are also piezoelectric. The present detectors are hard mounted to the package but do not give any false alarms unless the package is subjected to a sharp hit. A photograph of the pyroelectric detector module is shown in figure 2. Each detector module has an open collector output which permits a number of modules to be connected in parallel. Connecting detector modules in parallel reduces scintillation effects and permits customizing the hit area on target vehicles. The threshold level of each module is normally set to the same value (50 micro W) but each module is individually adjustable.

Fig. 2. Lithium Tantalate Pyroelectric Detector Module.

The CO_2 laser shown in figure 3 with its beam expander generates a gaussian beam with a 50% intensity width of 0.6 mr. A plot of the profile of the measured data is shown in figure 4.

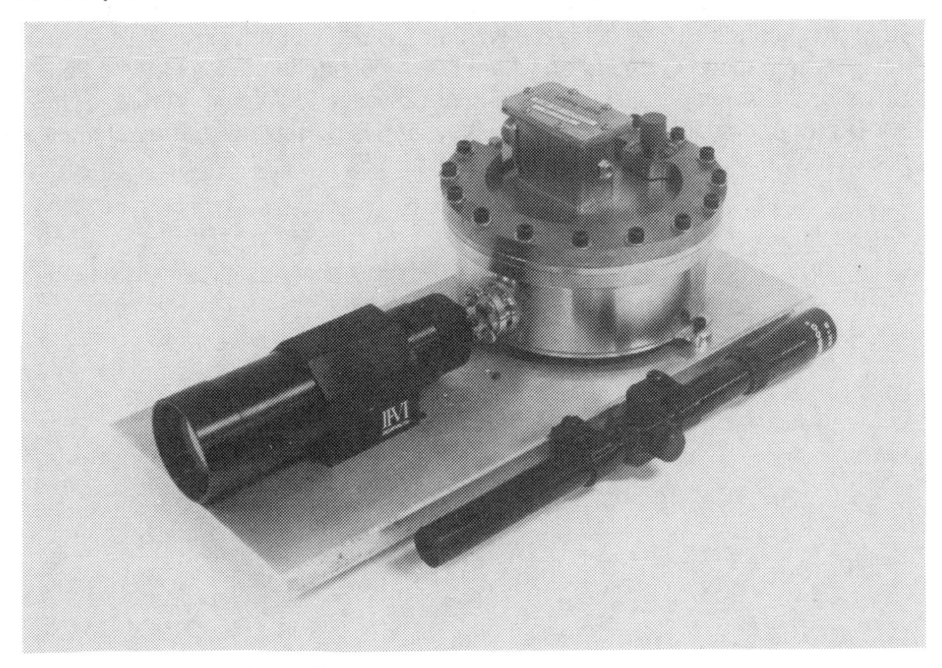

Fig. 3. CO_2 laser with beam expander optics.

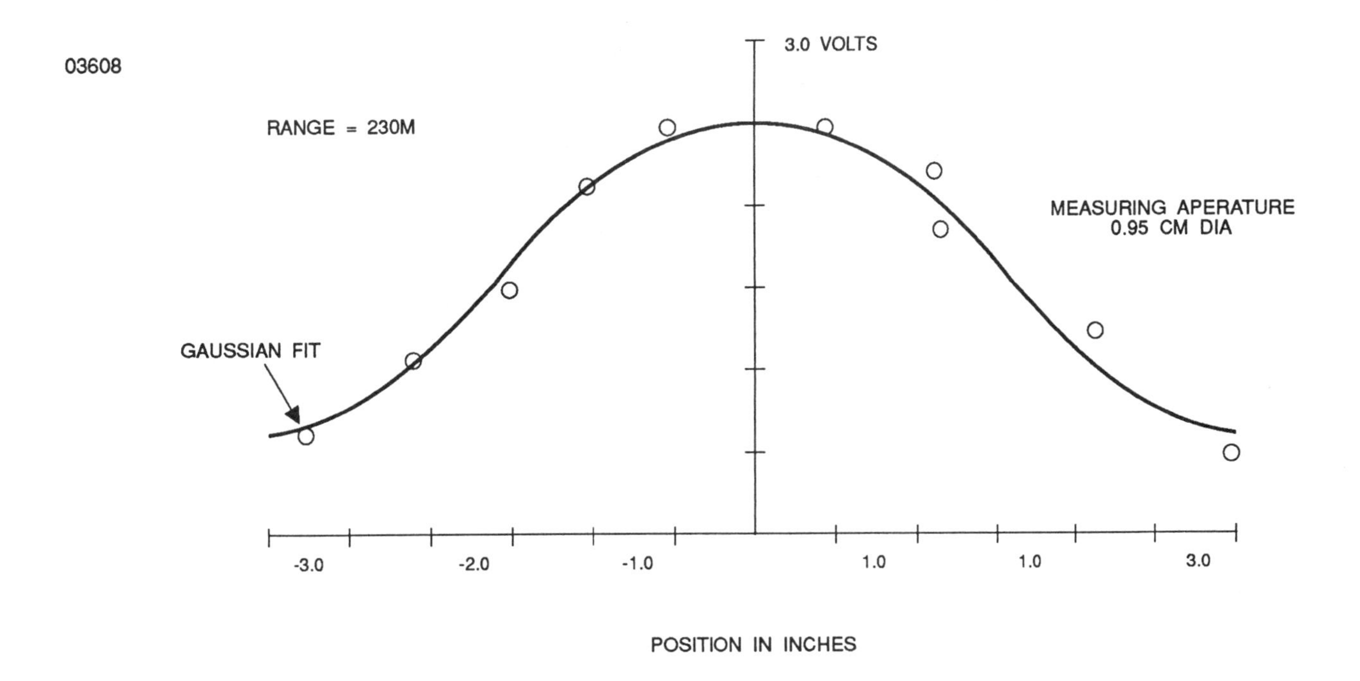

Fig. 4. Laser Beam Profile.

The laser is a completely sealed unit and has been operating for 9 months with no noticeable drop in its output power. The laser driver is micro processor controlled which facilitates generating codes and changing the laser output pulse width. The laser pulse width has been set at 300 micro seconds for initial testing, however pulse widths as low as 50 micro seconds can be achieved with the laser driver circuitry. This laser can operate cw with cooling. In the training system mode of operation where laser messages are less than 0.25 s there is no need for cooling either the laser or the rf supply. The laser output power is 25 W with 400 W of rf input power. The rf supply requires 28 V DC for operation. The present test system is portable with the batteries and microprocessor electronics contained in the laser controller unit shown in figure 5.

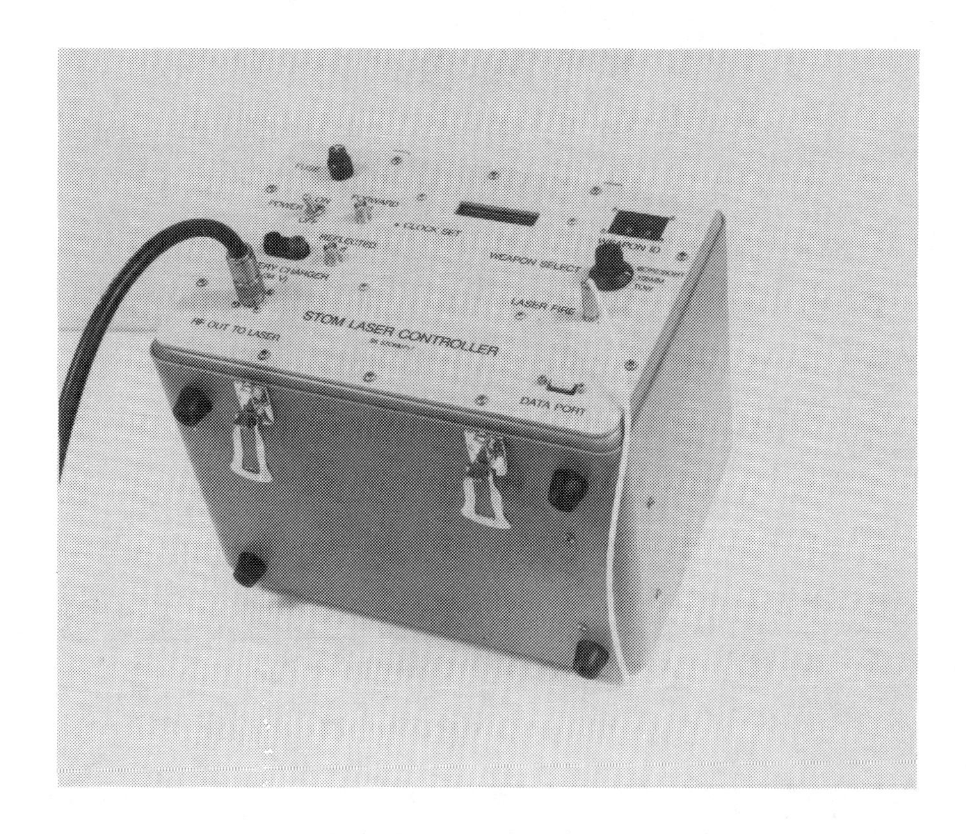

Fig. 5. Laser Controller.

2. DUST TRANSMISSION TESTS

2.1 Importance of operating in a dust environment

The single area involved in assessing the feasibility of a STOM system which is not well documented is that of attenuation due to dust. Unlike smoke attenuation, which is well documented, attenuation due to dust has not been studied in great detail. Most of the literature that does exist is based on theoretical estimates derived from Mie scattering theory. It is well known that during MILES training exercises the performances of the GaAs (900 nm) laser transmitters are degraded in the presence of dust clouds generated by winds or vehicles. Laser hits cannot be obtained even when the target is visible with the unaided eye with dust present. One important goal of the STOM program was the determination of the quantitative effects of dust in battlefield training.

2.2 Qualitative attenuation principles

While very little quantitative information is available, the following well-established qualitative principles are valid:

- Attenuation due to particles of diameter "d" will tend to be at or near its maximum at a wavelength λ which is approximately equal to d.
- When $d \ll \lambda$, attenuation will be significantly reduced relative to the maximum value.
- Based upon Stokes theory, large dust particles will fall back to earth very much faster than small dust particles.

In the absence of convection effects due to wind, particles approximately 10 micrometers in diameter (CO_2 wavelength) will fall from a height of a few feet in a time interval on the order of a few seconds. However, a particle having a diameter of 1 micrometer or less (near MILES laser wavelength) will take hundreds of seconds to fall to earth.

As a consequence, MILES, which operates at $\lambda = 0.904$ micrometers, will have its signal seriously attenuated for long periods of time, due to the small suspended dust particles, while a 10.6 micrometer based STOM system should experience significant attenuation for only a few seconds, followed by a longer interval in which the attenuation due to the smaller dust particles is finite but significantly reduced. It is precisely during this time interval that the major potential advantage of a STOM system should occur. The FLIR, which operates in the 8-14 micrometer wavelength, will be affected in much the same way as the CO_2 laser. As dust clears to where the FLIR can see, the CO_2 laser will also be able to see. The soldier will then be able to fire at his target as he would in combat.

2.3 Field test

On September 12, 1989, field tests to evaluate the effects of dust were conducted at the National Training Center (NTC) at Fort Irwin, California. A STOM channel, a MILES channel, and a visual channel were set up so that all three optical lines-of-sight were essentially coincident and equally affected by any intervening dust. The STOM channel detector output was recorded on a band of the strip-chart recorder, the MILES channel detector output on a second band of the same strip-chart recorder, and the visual channel on a VCR. By using a clock on the strip-chart recorder and a second clock within the field-of-view of the VCR, it was possible to correlate the data for the 10.6 micrometer (STOM) 0.9 micrometer (MILES) and visible channels simultaneously.

2.3.1 STOM channel

A STOM 10.6 micrometer CO_2 laser was mounted on a tripod and aimed at a Lithium Tantalate detector 100 meters away. The laser was boresighted, and the detector was tested to ensure that it was successfully receiving 10.6 micrometers of radiation.

2.3.2 MILES channel

A standard MILES transmitter was mounted on the tripod approximately one foot from the STOM laser and aimed at a standard MILES detector located 100 meters away within one foot of the STOM detector. The laser was boresighted, and the detector was tested to ensure that it was successfully receiving the 0.9 micrometer radiation.

2.3.3 Visual channel

A TV camera/recorder was placed next to the detectors and mounted on a tripod aimed at a large resolution chart placed next to the transmitters. The zoom lens was adjusted until the resolution chart filled at least half of the Field-Of-View (FOV). The clock was set up in the near foreground so that the sweep second hand was also clearly visible. Both the resolution chart and the clock were clearly in focus. In this way, the precise duration of the dust obscuration was recorded. The dust cloud was produced beyond the clock, but in front of the acuity target/detector location.

2.3.4 Strip-chart recorder

The strip-chart recorder was positioned adjacent to the 10.6 and 0.9 micrometer detectors next to the camera. In this way, 10.6 micrometers, 0.9 micrometer, and visible channel results could all be recorded simultaneously.

2.3.5 Dust generator

A vehicle was used to generate sustained dust levels by driving back and forth across the optical line-of-sight numerous times in an area of the desert that is ideally suited to the generation of large quantities of dust.

2.5 Test results

There were three sequences from the data taken that were analyzed. They are discussed below as sequences A, B, and C. The strip-chart data is not shown here but is available for review if desired.

2.5.1 Sequence A

A particularly effective dust episode, designated "Sequence A," occurred from approximately 12 hours, 37 minutes, 56 seconds until approximately 12 hours, 38 minutes, 09 seconds. The MILES GaAs channel at 0.904 micrometers drops to zero for approximately 13 seconds. During this interval, the STOM CO_2 channel at 10.6 micrometers oscillates about a mean value signal level equal to approximately 48% of maximum, with minimum value of 12% and maximum value of 100%.

1. 88% of the time the signal level exceeds 30% of the maximum value.
2. 72% of the time the signal level exceeds 40% of the maximum value.
3. 50% of the time the signal level exceeds 48% of the maximum value.

Furthermore, based on the data for the MILES channel, throughout this time interval we can safely say that the MILES signal with dust present $S_{m,d}$ < 0.03, since

$$S_{m,d} = S_{m,c}\exp(-k_{m,d}C_d\,\delta X)$$

where,

$k_{m,d}$ = extinction coefficient for the MILES channel at 0.9 micrometer, due to dust.

$S_{m,c}$ = MILES signal when atmosphere is clear.

C_d = dust concentration - grams/m^3.

δx = thickness of the dust cloud - meters.

Based on what little available data exists for dust attenuation, we shall estimate

$$k_{m,d} = 0.26\ \text{m}^2/\text{gram}$$

thus

$$0.03 = 1.00 \times \exp(-0.26[C_d\delta X]_{min})$$

where the minimum value for the $[C_d\delta X]$ product is appropriate since $S_{m,d} < 0.03$. Thus

$$\exp(0.26[C_d\delta X]_{min}) = 33.3$$

or $[C_d\delta X]_{min} = (1/.026)\ \ln(33.3) = 13.5\ \text{grams/m}^2$.

This is significantly greater than the original STOM requirement that $[C_d\delta X]$ = 0.2 gram/m$_3$ x 10 meters = 2 gram/m^2. Even if our estimate of $k_{m,d}$ is incorrect by a factor of 2, the dust concentration during sequence A of the field test was well in excess of the system requirement. If we assume that δX was, indeed, approximately 10 meters, then

$$C_d = 13.5\ \text{gram/m}^2/10\ \text{m} = 1.35\ \text{gram/m}^3$$

This would be almost a factor of 7 greater dust concentration than specified.

We can use this data to determine the effective dust extinction coefficient for the 10.6 micrometers by solving the inverse problem. Specifically let us calculate what the MILES channel signal level should have been if k, C_d and δX were the specified values, find such a data point, determine the actual value of the CO_2 channel signal, and, knowing the product of $C_d\delta X$, calculate the effective dust extinction coefficient for the STOM system, $k_{s,d}$

Since:

$$S_{m,d} = S_{m,c}\ \exp(-0.26 \times 0.2 \times 10)$$

$$= 100 \times 0.59$$

$$S_{m,d} = 59\%$$

Inspecting the MILES channel, we find a point where $S_{m,d}$ = 59% at approximately 12 hours, 37 minutes and 54.5 seconds. At this point $S_{s,d}$ = 0.85. Other such points yield values of 0.90, 0.92, 0.87 and 0.85. Since these values are reasonably consistent, let us take the average value.

$$S_{s,d} = (0.85 + 0.90 + 0.92 + 0.87 + 0.85)/5$$

$$= 0.88$$

Now, solving Beer's law for the extinction coefficient $k_{s,d}$ for the STOM system, at 10.6 micrometers, where by definition the concentration - path length product for the smoke cannot have changed, then

$$0.88 = 1.00 \exp(-k_{s,d}^{2})$$

or $$k_{s,d} = \tfrac{1}{2} \ln(100/88) = 0.50 \times 0.128$$

or $$k_{s,d} = 0.064$$

This is one of the first estimates of the extinction coefficient of dust at 10.6 micrometers. It suggests that at CO_2 laser wavelengths the extinction coefficient due to dust (at least the dust encountered in these tests) is significantly lower than the corresponding value at GaAs laser wavelengths. In this case

$$k_{s,d} + k_{m,d} = 0.064/0.26 = 0.25$$

To insure that this conclusion is statistically significant, it is important to check for additional confirming events.

2.5.2 Sequence B

Sequence B occurred between 12 hours, 38 minutes 19 seconds and 12 hours, 36 seconds. Here again the MILES channel is very low (i.e., near zero), while the STOM channel fluctuates significantly but shows signals averaging near 50% again.

1. 95% of the time the STOM signal level exceeds 10% of the maximum value.

2. 85% of the time the STOM signal level exceeds 20% of the maximum value.

3. 75% of the time the STOM signal level exceeds 30% of the maximum level.

4. 61% of the time the STOM signal level exceeds 40% of the maximum value.

5. 45% of the time the STOM signal level exceeds 50% of the maximum value.

To simplify these results, let us define $\phi(S)$ as the percent of the time the STOM signal exceeds the fractional signal level S, defined as

$$S \equiv S_{s,d} / S_{max} \quad \text{(in percent)}$$

The results of data for Sequences A, B, and C are plotted in fig. 4.

Although the trend is similar for both Sequence A and Sequence B data, there is a systematic displacement of 5 to 10 percentage points in the value of ϕ. Returning to Sequence A, we can also note that $\phi = 100\%$ for $S = 10\%$, and $\phi = 99\%$ for $S = 20\%$.

2.5.3 Sequence C

The results of Sequence C are :

ϕ	=	100% for S	= 10%
	=	94% for S	= 20%
	=	85% for S	= 30%
	=	70% for S	= 40%
	=	51% for S	= 50%

The data for Sequence C again shows a trend simlar to those of Sequences A and B.

2.6 Conclusions

To a first approximation, it would appear that the effective extinction coefficient at 10.6 micrometers, for the specific dust/particle distribution occurring in these tests, is approximately 0.06 m^2/gram.

Data gathered during this field test indicate that attenuation at the STOM/CO_2 wavelength of 10.6 micrometers is significantly reduced relative to the MILES/GaAs wavelength of 0.9 micrometers, as well as the visual spectrum.

In essence, this data indicates that approximately 50% of the time the STOM system signals will exceed 50% of their maximum level, during dust conditions which have reduced MILES system signals to less than 3% of their maximum level and which have totally obliterated the view of the target.

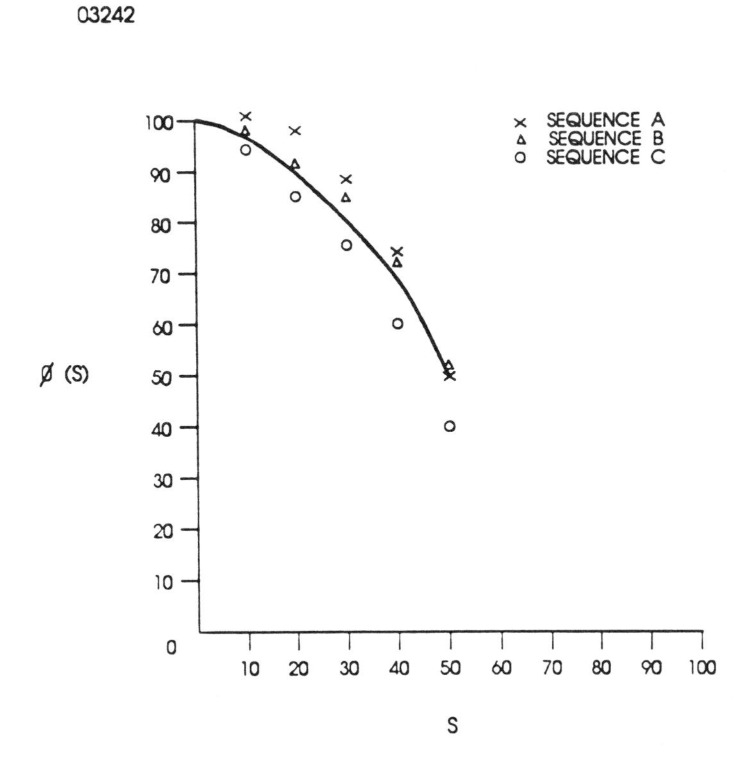

Fig. 6. Plot of sequence data.

3. RANGE TESTS

3.1 Hit profiles vs range

The laser transmitter shown in figure 3 was mounted on a tripod and an additional 32 power sighting scope (not shown) was attached to the unit in order to perform long range testing. The laser controller's output rf cable was then attached to the laser. A 105 mm main gun code was chosen for the tests. The pyroelectric detector module shown in figure 2 was connected to Microprocessor Decoder/Strobe Assembly (MDSA) shown in figure 7. The beam diameter was measured as a function of range. The results are given in table 1 below.

TABLE 1. BEAM DIAMETER AS A FUNCTION OF RANGE

RANGE(km)	DIAMETER(m)
0.5	1
1	2
2	2
3	2
4	2
5	2
6	2

Only one detector module was used in these range tests. Additional testing will be run in the near future to measure the effects of using multiple detectors and determining the actual 90% hit probability zone both in the vertical and horizontal position. The maximum range for the present system is estimated to be 8 km. Additional testing will determine the centerline hit probability as a function of range. The MDSA is being programmed to count the words received during each firing and display this number on its liquid crystal display. This function is required to be completed before additional testing is started. The MDSA also contains a battery backed up clock, ram, and an RS-232 port for dumping stored events. The MDSA can store up to 1,000 events such as weapon type, time, and player ID.

Fig 7. Microprocessor decoder/strobe assembly.

Carbon dioxide eyesafe laser range finders

Richard Powell
Barry Berdanier
Jim McKay

Texas Instruments, Dallas, Texas

ABSTRACT

Eyesafe military laser range finder systems that incorporate carbon dioxide lasers operating at 10.59 microns have been successfully developed and are currently in production for both the U.S. and foreign military services. The development of carbon dioxide laser rangefinders for Fire Control applications has provided high performance,eyesafe capability to both heavy combat vehicles and air defense platforms. The distinct wavelength compatibility provided by a long wavelength laser in conjunction with long wavelength Forward Looking Infrared (FLIR) detection systems positions the CO_2 laser range finder system as unique in the ability to provide ranging capability to FLIR recognizable targets and eye safety. First order modelling of expected performance versus system design parameters has provided a basis for understanding the key performance factors and their relationship to necessary design trade-offs. The successful implementation of the laser range finder design required the development of an array of CO_2 laser system components that provide both the transmitted laser pulse and the ability to detect the target reflected return. A modular design approach to the CO_2 laser system components has led to several successful programs that incorporate identical key technologies thereby reducing the overall cost of all CO_2 laser range finder programs.

1. INTRODUCTION

The advantages of eyesafe laser systems over non-eyesafe laser systems for military applications are most evident in the increased training opportunities that eyesafe lasers provide to military personnel. Non-eyesafe laser systems are severely restricted in that only a limited number of government facilities for realistic field training exist. An additional and often overlooked limitation to non-eyesafe systems is that due to the relatively limited amount of field testing opportunities for hazardous devices, the possibility of fielding a production design that may still have minor flaws is greater than with an eyesafe system that can be tested more thoroughly at the early stages of development. The ability to detect and correct design flaws early in a laser product's life cycle is obviously advantageous to the manufacturer as well as the customer in keeping costs down as well as meeting the technical requirements for the system.

Prior to the introduction of solid state eyesafe laser devices, the only eyesafe laser technology was based on the carbon dioxide (CO_2) laser. The existing CO_2 laser range finder technology has many benefits as compared with the emerging solid state technologies including development maturity and the ability to perform range finder functions as well as advanced active/passive sensor functions that increase system operational capability.

2. CARBON DIOXIDE LRF ADVANTAGES

The technical/operational advantages of a CO_2 LRF include wavelength compatibility of the range finding device with 8-12 micron night vision devices typically referred to as FLIRs (Forward Looking Infra Red) and the ability to penetrate common smokes that are found on the battlefield. The operational advantages related to advanced active/passive sensor applications is deferred to the following section on applications.

2.1 FLIR compatibility

Due to the long wavelength emission (10.6 micron) of the CO_2 laser occurring within the FLIR 8-12 micron band a greater degree of operational compatibility exists than between FLIRs and short wavelength lasers. Shorter wavelength lasers such as Neodymium (1.06 micron) and Erbium/Raman shifted Neodymium (1.54 micron) are well outside the FLIR's operating wavelength region resulting in situations where the target can be seen in the FLIR but not ranged on by shorter wavelength laser systems. The ability of FLIR systems to 'see' through conditions of poor visibility is well matched with a CO_2 laser systems ability to range through the same poor visibility conditions. This compatibility with respect to operating wavelength can also be advantageous when both a FLIR device and a range finder are required to meet the overall system requirements. System designers can leverage this advantage by combining laser and FLIR optical paths to reduce system physical size and cost. An example of this type of compatibility is described in the applications section on main battle tank combat vehicles.

2.2 Smoke performance

The ability of long wavelength radiation to penetrate obscurants is well understood. The CO_2 LRF systems have an advantage over shorter wavelength lasers through the ability to penetrate common smoke and dust. In the case of comparing eyesafe laser wavelengths the CO_2 wavelength is approximately 7x the solid state technology 1.54 micron wavelength and consequently does perform better versus common smokes. Specialty smokes that obscure both the long infra red (IR) and the solid state 1.54 micron will unfortunately prohibit ranging with either wavelength system as well as detection of the intended target with visible and IR optical sighting systems. If the target can not be seen then the question of which laser wavelength is best is a moot point. Fortunately, the CO_2 laser system and the FLIR do very well in the most common smokes such as diesel oil and fog oil which are the most widely used.

3. CARBON DIOXIDE LRF APPLICATIONS

The eyesafe CO_2 LRF has been applied to military fire control applications to determine both target range and range rate.They are also being used as a weapon guidance laser communication system to steer anti-tank projectiles to the target. Advanced applications include both chemical agent detection systems and combination active/passive IR sensors.

3.1 Air defense fire control

The air defense application of eyesafe LRF's provides accurate range and range rate data to the vehicle fire control system. The first production eyesafe fire control LRF to enter the U.S. Army inventory was the CO_2 laser range finder for the Avenger Pedestal Mounted Stinger (PMS) air defense fire control application. A summary of technical specifications for the Avenger LRF is shown in Table 1. The LRF is physically located on the Avenger Stinger missile launch platform and is bore sighted to the fire control FLIR system. The LRF, Figure 1, is a dual aperture LRF that receives fire commands and vehicle power as inputs and provides range data and built in test data to the fire control computer as outputs.

TABLE 1

AVENGER LRF
TECHNICAL SPECIFICATIONS

TYPE	EYESAFE CO_2
PULSE RATE	10 HZ
COOLING	LINEAR COOLER
COOLDOWN	< 5 MINUTES
WEIGHT	< 50 LBS
SIZE	7" X 13" X 20"
RANGE WINDOW	400 - 9990 METERS
ACCURACY	+/- 10 METERS
INPUTS	20-30 VDC, 100 WATTS, FIRE COMMAND
OUTPUTS	RANGE DATA, BIT DATA
ENVIRONMENTAL	MIL-STD-810D

A simplified block diagram of the Avenger LRF is shown in Figure 2. Vehicle input power is filtered and provided to the laser electronics assembly which powers the pulse forming network (PFN) and fan module. The PFN provides the high voltage, high current electrical pulse necessary to power the transversely excited atmospheric (TEA) laser. The fan module is an integral part of the laser transmitter and is described in more detail as is the electronics in later sections. The optical output of the laser is passed through a 2.2x refractive beam expander where the output power is monitored and a signal provided to the range computer to start the range clock. Then a 5x reflective beam expander further reduces the beam divergence with the output window providing environmental protection to the internal volume of the LRF. After reflection from the target down range the laser energy is collected by the afocal optics and imaged onto a cryogenically cooled Mercury Cadmium Telluride photo conductive detector. The resulting electrical signal is processed by the receiver circuit card assembly (CCA) and provided to the range CCA to stop the range counter. The low voltage power supply (LVPS) CCA and the cooler power supply meet regulated power needs and the logic CCA provides control functions. The range data is provided on the signal output connector located adjacent to the LRF test connector.

An additional air defense application of CO_2 laser technology for range and range rate data is the Light Armored Vehicle - Air Defense (LAV-AD) program for the U.S. Marine Corps. The LAV-AD LRF, is a coaxial system using many of the same components as the Avenger LRF and is currently undergoing government testing.

3.2 Main battle tank combat vehicles

The first production application of a an eyesafe CO_2 LRF to military vehicle fire control was the development of a range finder that takes full advantage of the wavelength compatibility between the FLIR and the laser. The Gunner's Primary Tank Thermal Sight (GPTTS) developed for the South Korean military's main battle tank (K1) employs a CO_2 laser transmitter and the existing FLIR optics, detectors and cryogenic cooling to detect target reflected laser range returns. The advantages of this approach include a reduced number of expensive long wave IR components such as optics, detectors and cryogenic coolers for the fire control system as a whole as compared to the fire control system that requires dedicated CO_2 range finder receiver modules (optics,detectors,etc).

Figure 1, Avenger LRF

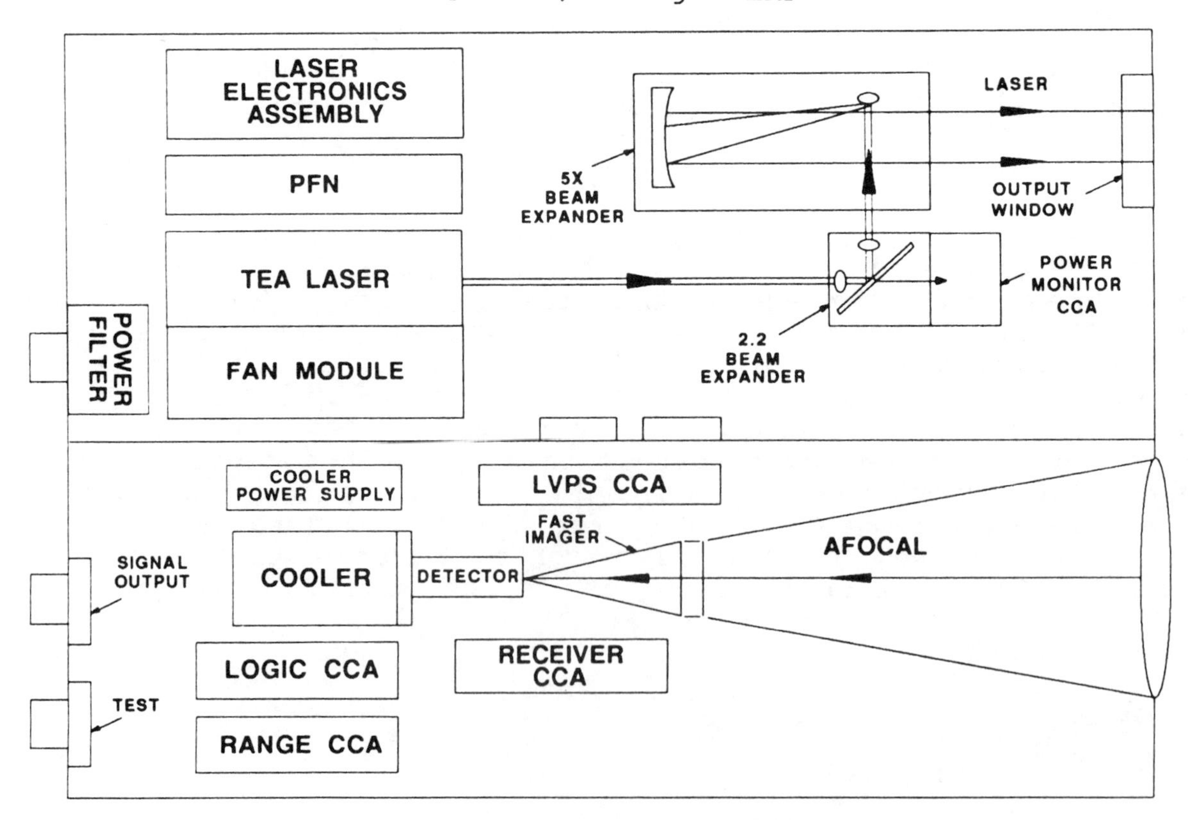

Figure 2, LRF Block Diagram

3.3 Weapon guidance

The ability of long wavelength CO_2 laser radiation to effectively transmit through obscurants as described earlier is a key element in the guidance of advanced anti-tank weapons. The development of kinetic energy projectiles that can be guided after launch using a pulsed CO_2 laser takes advantage of the obscurant transmission capability by communicating data through the missile exhaust plume to increase the probability of hitting the target. A U.S. government program, Line Of Sight Anti Tank (LOSAT), to demonstrate the overall effectiveness of this type of weapon system is currently in government testing. The CO_2 laser system provides the ability to communicate with the missile through the exhaust plume as well as providing conventional target range finding data to the fire control system.

3.4 Chemical agent detection

An advanced application of CO_2 laser range finder technology is the detection of hazardous chemicals. The ability of the CO_2 laser to transmit radiation at many wavelengths where hazardous chemical agents exhibit identifying absorption spectra allows the detection and identification of many hazardous airborne chemical agents. Modifications to the LRF transmitter and receiver include addition of a wavelength tuning scanner to the TEA laser, additional receiver electronics and automatic data processing capability. A development program sponsored by the U.S Army Chemical Research, Development and Engineering Center, Aberdeen Proving Ground, Maryland addresses the approach of using CO_2 laser technology to provide stand-off detection capability of hazardous chemicals.

3.5 Combination active/passive IR sensors

The existance of the common IR waveband for CO_2 lasers and 8-12 micron passive IR sensors provides many opportunities for combining multiple capabilities of the vehicle level electro-optical system into singular integrated systems. Many of the features of a CO_2 laser receiver (optics,detectors,cryo-coolers,etc) are also required for passive infrared sensor (PIRS) applications. The continued development of staring 2 dimensional focal plane arrays provides the opportunity to combine LRF functions with other functions reducing associated size, weight and cost. The availability of a non-scanning 8-12 micron IR optical port that can be aimed at targets of interest as well as determine range to the target is a significant advantage to the system designer tasked with integrating both current and advanced active/passive sensors into a complete vehicle fire control system.

4. PERFORMANCE MODELLING.

A successful laser rangefinder design/development/production effort must meet the performance expectations of the customer as well as the cost and reliability goals of the program. The correct determination of the specific minimum performance requirements necessary to meet the customer's overall performance expectations are critical to the success of any program. Typically, these expectations must be translated into a set of testable requirements. Once the requirements have been determined the suitability of a particular device must be evaluated with respect those requirements and the performance of the LRF predicted at other conditions/situations of interest.

4.1 Requirements definition

Addressing the maximum range performance requirement of a laser rangefinder system as an example, a typical requirement will be described as a minimum probability of ranging to a particular target, at a minimum acceptable distance under weather conditions that are equal or better than specified. An additional factor that is often not provided is the amount of pointing error between the intended target and the laser system. Since the maximum range requirements will typically describe a worst case type situation, the opportunity for actual testing of the LRF versus those worst case conditions can be extremely limited or totally impossible. Consequently, an alternative test method must be used to simulate the required conditions and determine whether or not the LRF system meets the requirement.

The alternative test most commonly used is called an extinction test[1]. The extinction test method requires the insertion of attenuation filters in the LRF to test target optical path to simulate ranges beyond the test range and to simulate the required atmospheric transmission effects. To determine the amount of filtering that must be inserted and still provide a specified probability of ranging to the test target a ratio analysis must be completed.The ratio analysis identifies the differences between the required worst case condition and the extinction test condition. The differences between the test condition and the worst case required performance conditioned can be divided into the following categories:

I. Range factor = the difference between the required target range and the test target range

II. Transmission factor = the difference between the required atmospheric transmission and the test condition transmission

III. Reflectance factor = the difference between the required target reflectivity and the test target refectivity

IV. Target size factor = accounts for the loss of reflected energy due to the size of the required target at the required range

V. Pointing factor = accounts for the loss of reflected energy due to pointing error between target and LRF

VI. Probability factor = accounts for the difference in required probability of successful ranging (typically 99%) versus the typical extinction test probability (typically 50%)

Once all of these factors are identified the amount of filtering required to simulate the required performance can be calculated. This listing of factors is a first order model that allows the LRF test requirements with respect to the desired maximum range performance to be defined.

4.2 Performance predictions

Once an extinction test requirement for LRF maximum range performance has been defined based on the customer's required range performance, predictions and/or testing of the ability of the LRF's to meet that extinction test requirement can be done. When the LRF has been predicted and/or tested to meet the extinction test requirement then the system designer's task will usually be to predict range performance versus different targets, different weather conditions, etc. than the original customer performance requirements. A first order range performance model derived from the general equation[2] for received radiant power at the LRF detector is shown below:

$$(SNR) \times (NEP) = \frac{M_C \times P_T \times A_R \times T_R \times R_T}{\pi \times R_M^2} \times \exp^{-2AR_K}$$

NEP = NOISE EQUIVALENT POWER OF THE LRF RECEIVER (WATTS)

SNR = MINIMUM SIGNAL TO NOISE RATIO REQUIRED TO ATTAIN A SPECIFIED PROBABILITY OF DETECTION AND FALSE ALARM RATE

M_C = FRACTION OF ENERGY INTERCEPTED BY THE TARGET AND WITHIN THE LRF RECEIVER FIELD OF VIEW

P_T = TRANSMITTED LASER POWER DIRECTED TOWARD TARGET AFTER TRANSMISSION LOSSES DUE TO TRANSMITTER OPTICS (WATTS)

A_R = COLLECTING AREA OF THE LRF RECEIVER OPTICS (SQUARE METERS)

T_R = TRANSMISSION OF THE RECEIVER OPTICS

R_T = TARGET REFLECTANCE

R_M = RANGE TO TARGET (METERS)

A = (ALPHA) = ATMOSPHERIC ATTENUATION COEFFICIENT

R_K = RANGE TO TARGET (KILOMETERS)

A computer based formulation of this equation to predict range performance for both staring LRF receiver systems and LRF's that use the scanning FLIR detectors as the LRF receiver is required to efficiently predict range performance for CO_2 laser range finders described previously.

5. LASER ELECTRONICS

5.1 Control/Interface Electronics

The control/interface electronics consist of the Logic which includes both the system timing and interface functions and the Range Counter which calculates the range to the target. The Range Counter provides the target range to the fire-control computer in a binary-coded-decimal (BCD) format with a 10 meter least significant bit (LSB). The range accuracy provided in the existing systems is 10 meters (1σ), but a 5 meter LSB can be added giving a 5 meter (1σ) range accuracy when used with the appropriate Receiver Electronics. The Logic and Range Counter functions are two separate printed circuit boards in the existing systems but could easily be combined into one for a new design.

There is also a Low Voltage Power Supply which provides power for the Logic and Range Counter electronics as well as the Receiver Electronics. The Low Voltage Power Supply operates from standard vehicle power of 18 VDC to 32 VDC and requires a typical input power of 20 Watts.

5.2 Receiver

The Receiver takes the target signal from the detector, processes it, detects it, and provides a differential "stop pulse" to the Range Counter. The differential interface between the Range Counter and the Receiver not only provides electrical isolation but also allows the Range Counter to be remoted from the Receiver.

The Receiver utilizes a photo-conductive (PC) HgCdTe detector. The typical detector has a time constant of .5 microseconds giving an electrical bandwidth of only 300 KHz, Figure 3. The dominant noise source in the detector, generation-recombination noise, rolls off by the same time constant as the signal, and, as a result, the detectivity [$D^*(10.6)$] stays flat until the generation-recombination noise drops below the thermal or Johnson noise of the cooled detector which occurs out past 8 MHz.

The Receiver combines the PC HgCdTe detector with a frequency compensation circuit and low noise amplifier to give a circuit providing an NEP of less than 2.5 nW, a target separation of less than 30 meters, and the risetime needed to allow the system to easily meet the 10 meters (1σ) range accuracy requirement.

5.3 Transmitter electronics

There are three electrical assemblies that power and monitor the transmitter, the High Voltage Power Supply (HVPS), the pulse forming network (PFN), and the Power Monitor.

The HVPS takes vehicle power and generates a 2.5 KV square wave which drives the PFN. The PFN has an internal voltage multiplier that converts the 2.5 KV into a 18 KVDC charge on an energy storage capacitor. The laser transmitter is fired by applying a trigger pulse to the PFN spark gap, discharging the energy storage capacitor into the laser cavity.

The Power Monitor samples a small portion of the laser beam and provides a pulse which is used as both the "start pulse" command to activate the counters on the Range Counter and as a relative measurement of laser output for the built-in-test (BIT) function. The Power Monitor consists of a pyro-electric detector coupled with a transimpedance amplifier which includes a capacitance neutralizing circuit for offsetting the detector's capacitance.

The power consumption of the transmitter electronics are a function of the repetition rate of the laser. The transmitter electronics in a typical Avenger system, for example, running 10 pps would pull 32 Watts.

6. Laser transmitter description

The CO_2 laser transmitters currently built for laser range finder applications are of two similar designs, one intended for 1 pulse per second (PPS) operation and the other for higher repetition rates. The design, Figure 4, utilizes an aluminum housing, Chang profiled electrodes and ultraviolet (UV) preionization. The discharge volume is approximately 8 cubic centimeters (CC). The optical resonator is a simple two mirror design consisting of a flat, zinc selenide outcoupler and a concave, diamond machined copper total reflector. The mounting surfaces of the laser housing and resonator mirrors are designed such that very tight machine tolerances can be achieved thus permitting the optical resonator to be assembled without adjustment. Resonator alignment is maintained over the full range of military equipment environmental conditions.

Figure 5 shows the higher repetition rate design. It is identical to the previously described design except that a transverse fan is added to increase gas flow through the discharge region. The fan motor is magnetically coupled to the fan impeller.

The lasers are typically operated with a 30% CO_2 gas mixture at a total pressure of approximately one atmosphere. The PFN which supplies electrical energy to the transmitter stores 1.2 joules per pulse. Typical voltage and current curves are shown in Figure 6.

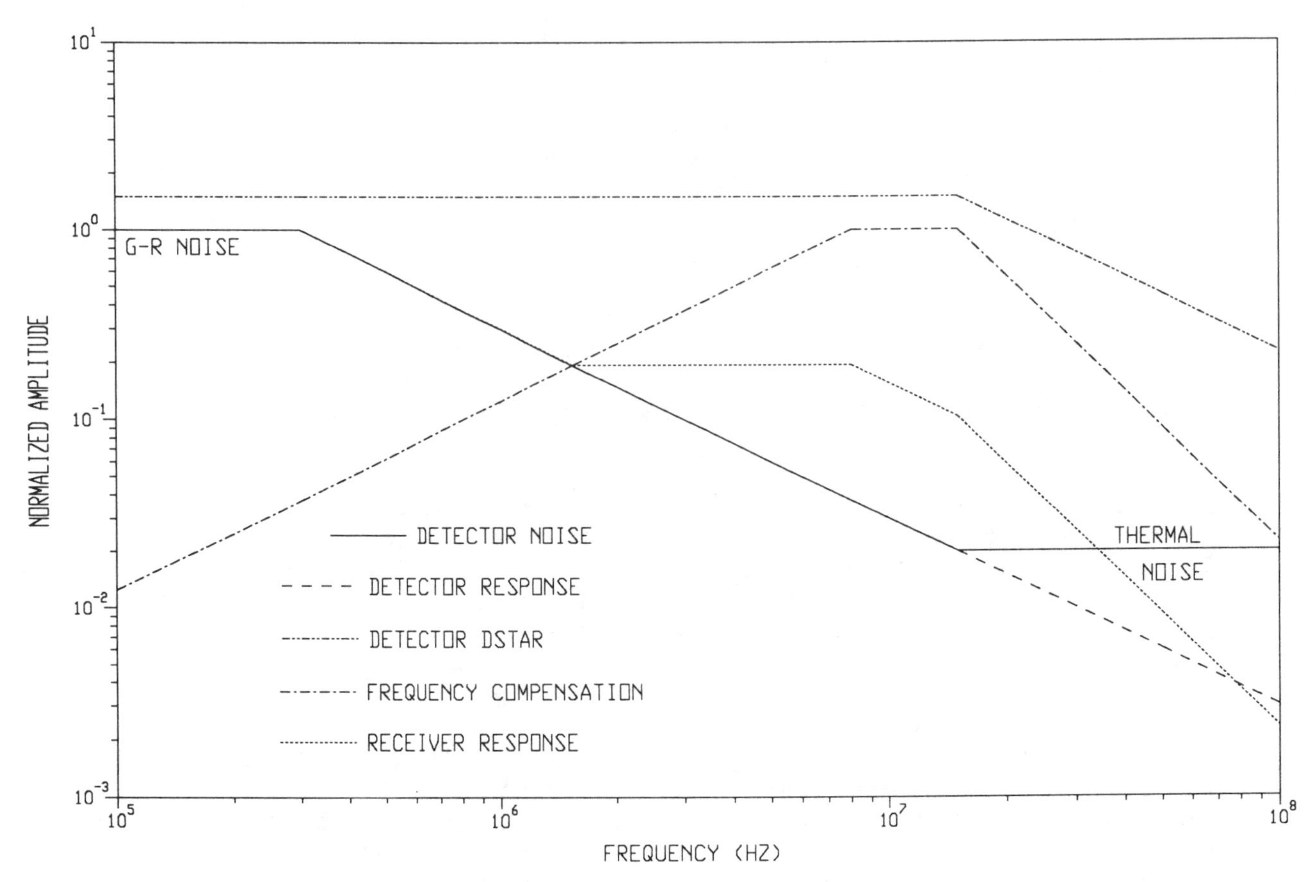

Figure 3, LRF Receiver Frequency Domain

Figure 4, Laser Transmitter, Low Pulse Repetition Rate

Figure 5, Laser Transmitter, Medium Pulse Repetition Rate

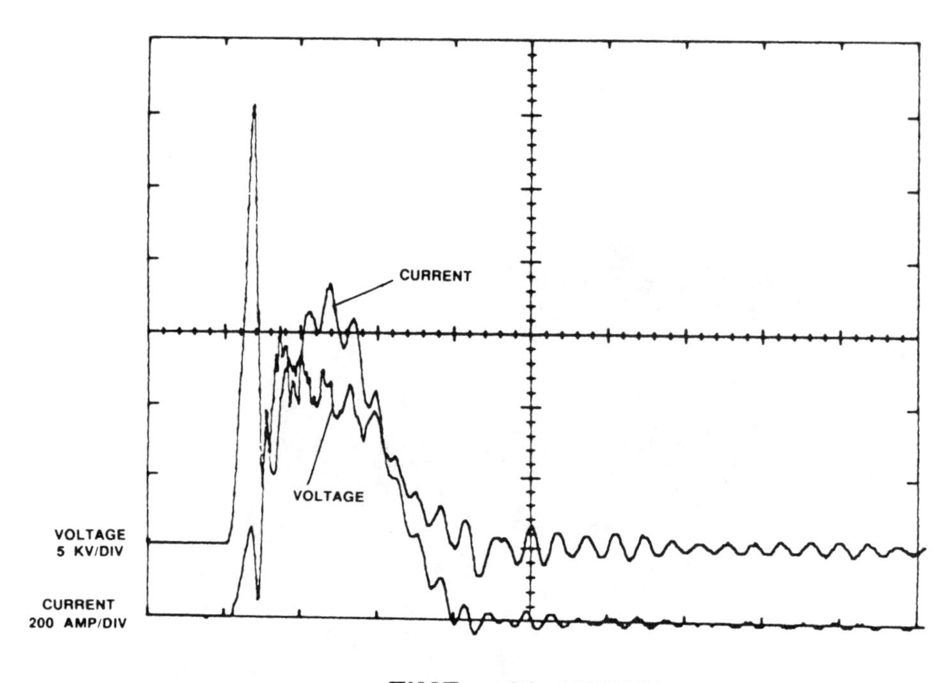

Figure 6, Laser Transmitter Voltage and Current Curves

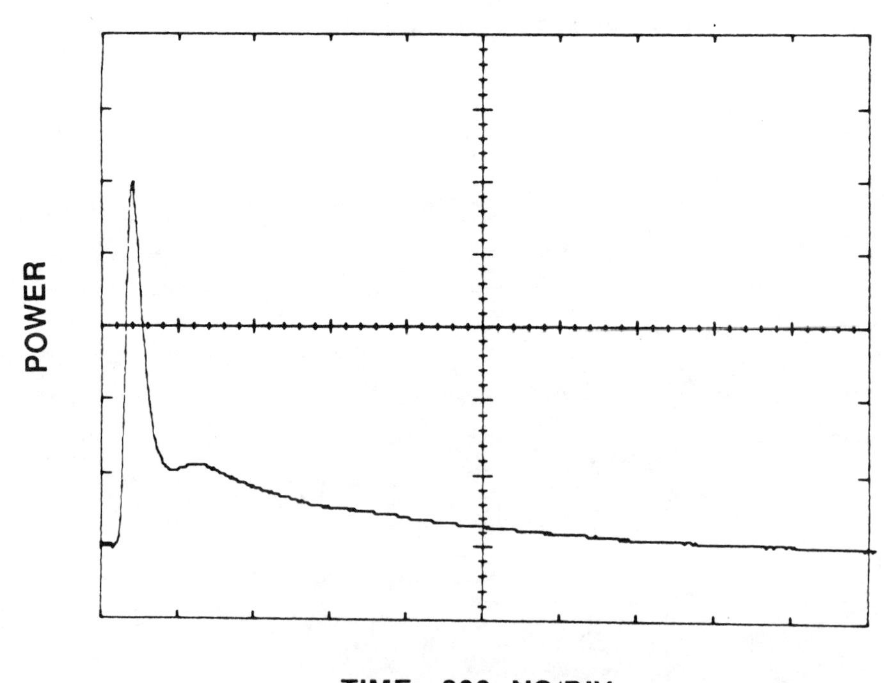

Figure 7, Laser Transmitter Output Pulse

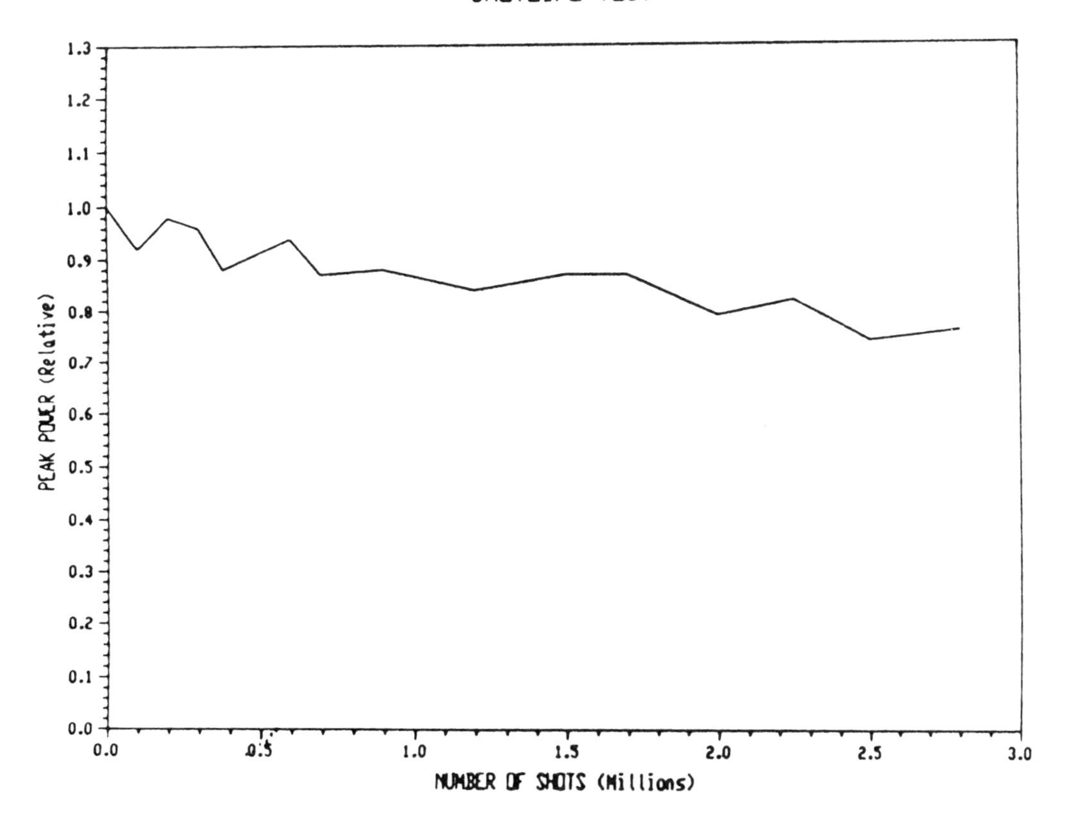

Figure 8, Laser Transmitter Shotlife Test

6.1 Performance characteristics

A laser transmitter output pulse is shown in Figure 7 and is characterized by a gain switched spike and subsequent tail. The peak power of the spike is typically 450 kilowatts (KW), the total pulse energy is 100 millijoules (mJ) and the Full Width Half Maximum (FWHM) pulse width is 60 nanoseconds (nS).

No effort is made to control the transverse mode structure of the output beam and consequently the laser exhibits a low order multimode character with a hot spot at the center containing most of the total peak power. The resulting far field beam divergence is 7 milliradians (mR).

The performance characteristics are maintained throughout typical military environmental testing. Laser transmitters have been operated at temperature extremes of -32 to +60 degrees Centigrade and have performed well after subjection to various vibration test profiles as well as shock levels as high as 200 g's.

6.2 Shotlife performance

Shotlife tests, Figure 8, have been run to greater than 2.5 million shots while monitoring a stable discharge throughout the test. The tests are run 6-8 hours per day continuously at 10 PPS. The peak power shows a fairly linear decay of 25% at 2.5 million shots. The power decay is due entirely to gas degradation since regassing of the laser following the test restores the peak power to the original value.

7. SUMMARY

The development of eyesafe CO_2 laser range finders for military applications has successfully transitioned from the prototype or development model stage to multiple production products based on a modular design approach. The advanced sensor integration application concepts coupled with the ever present need to reduce size, weight and cost of military electro-optical systems assure the continued development of competitive CO_2 eyesafe laser systems.

8. REFERENCES

1. M.L. Stitch, "Laser Rangefinding", Laser Handbook, Vol.II, pp. 1787-1788, North Holland Publishing, Amsterdam, New York, Oxford, 1972.

2. W.L. Wolfe and G.J. Zissis, Infrared Handbook, pp. 23-6, Environmental Research Institute of Michigan, 1978.

Highly efficient optical parametric oscillators

Larry R. Marshall, A.D. Hays, J. Kasinski, and R. Burnham

Fibertek, Inc., 510 Herndon Pky
Herndon, VA. 22070

ABSTRACT

An eyesafe source (1.61 µm) with 1.1 % wallplug efficiency, is demonstrated using a Nd:YAG pumped KTP optical parametric oscillator with peak-power conversion efficiencies of 70%. Joule-level scaling, kHz repetition-rates, and ns pulselengths are now accessible using this technology.

1. INTRODUCTION

The current interest in efficient lasers for applications involving coherent laser radar, remote sensing, and active imaging, has stimulated much interest in eyesafe lasers. The eyesafe requirement, essential for laser operation in populated areas, restricts the laser wavelength to a number of discrete bands in the infra-red spectral region. While such eyesafe outputs can be obtained directly from erbium and holmium lasers[1], or by Raman shifting in methane[2], we have developed a more efficient, all-solid-state approach using KTP-based optical parametric oscillators to frequency shift the output of highly-efficient diode-pumped Nd:YAG lasers. Optical parametric oscillators (OPO's) can provide an efficient and relatively simple means of frequency shifting; the output of a high-power pulsed solid-state laser is focussed into a non-linear crystal, and if the intensity of this pump laser is sufficiently high, the optical parametric process is initiated, generating output at different frequencies to that of the pump laser. Optical parametric oscillators employ a second-order non-linearity to convert photons at the pump frequency ω_p into photons at two 'new' frequencies, the so called "signal" ω_s and "idler" ω_i frequencies. Conservation of energy requires that the algebraic sum of the signal and idler frequencies be equal to the frequency of the pump.

In the past, OPO's have been of limited value for two reasons : Firstly, the required pump intensity is usually comparable with the damage threshold of the non-linear medium; since the efficiency of the OPO scales with pump intensity, efficiencies are usually restricted. Secondly, the output of the pump laser must be single transverse mode and of high optical quality; to achieve this, the efficiency of the pump laser is greatly reduced. This together with the low efficiency of the OPO, make the net efficiency of the resulting device quite low.

Recently, times these two problems have been overcome: Firstly, new non-linear crystals with high damage thresholds have been developed (for example KTP, BBO, KTA). Secondly, diode pumping has provided a new generation of efficient high-power solid state lasers with single transverse mode outputs.

2. THEORY

(2.1) Optical Parametric Process

The signal and idler outputs are generated by interaction between the intense pump and weak quantum noise at the signal and idler frequencies. This interaction occurs via the second order non-linearity of the crystal medium, and results in amplification of the signal and idler at the expense of the pump. The power gain G for the signal wave produced by a pump of intensity I_p in an optical parametric converter is given by[3]:

$$G=\cosh^2 \mathcal{G}L \qquad (1)$$

where $$G = \frac{2\omega_s\omega_i d_{eff}^2 I_p}{n_s n_i n_p \varepsilon_0 c^3},$$ (2)

and the effective length $\mathcal{L}$ is given by,

$$\mathcal{L} = \ell_w \mathrm{erf}(\ell\rho/w_p).$$ (3)

The walkoff angle is ρ, and w_p is the pump beam spot radius. For long crystals the effective length is given by the walkoff length, $\ell_w = \sqrt{\pi}\, w_p/2\rho$. The effective non-linear coefficient, d_{eff} is calculated from the second order nonlinear tensor of KTP.

Since the signal I_s and idler I_i intensities are extremely low, typically e^{-13} times lower than the pump intensity I_p, they must be greatly amplified before any significant depletion of the pump occurs. We therefore say that the optical parametric process demonstrates a threshold, and this threshold is typically defined as the point at which the pump depletion due to generation of signal and idler fields reaches 1 %[4] High pump intensities (typically 100 $MWcm^{-2}$) are required to exceed threshold for the optical parametric process, and this threshold intensity is often comparable with the damage threshold of the OPO crystal.

2.2 Optical Parametric Oscillators

In order to lower the threshold, the crystal is usually enclosed in a resonant cavity which provides feedback at the signal, and/or the idler frequencies. For stability reasons the cavity is usually resonant only at the signal (or idler) frequency. If both signal and idler are resonated, then all three fields must remain in phase over many passes through the cavity. This would require a single frequency, chirp-free laser. However, since the sum of the phases of the signal and idler waves is equal to that of the pump, it is possible to obtain a constant phase signal, if the phase of the idler matches itself to that of the pump. In this case a broadband (multi-longitudinal mode) pump can be just as effective as a single frequency pump, in driving the OPO[4]. When only one wave is resonated, the other is free to change its phase (and frequency) to match that of the pump. For these reasons, singly resonant oscillators (SROs) are used in most practical applications. The threshold for a singly resonant oscillator (SRO) is given, in the steady state case, by:

$$GL = [2\alpha\ell + \ln(1/\sqrt{R})],$$ (4)

where $2\alpha\ell$ is the round trip loss, and R is the cavity output mirror reflectivity. For an OPO pumped by a pulse of duration τ (intensity half-width at the 1/e point), the threshold pump intensity is[4]:

$$I_{th} = \frac{2.25}{kg_s\mathcal{L}^2}\left\{\frac{15L}{ct} + 2\alpha\ell + \ln\frac{1}{\sqrt{R}} + \ln 2\right\}^2.$$ (5)

where,

$$\kappa = \frac{2\omega_s\omega_i d_{eff}^2}{n_s n_i n_p e_0 c^3}$$ (6)

The effective non-linear coefficient d_{eff} is determined by phase matching considerations, as discussed below. The threshold of an OPO is therefore proportional to the square of the non-linear coefficient and effective crystal length. Threshold can be reduced in the SRO by allowing the pump to make a second pass through the OPO crystal. This is most easily done by reflecting the pump off the output coupler of the OPO . The threshold pump intensity I_{th} for a singly-resonant OPO with reflection of the pump from the OPO output coupler, is given by[4]

$$I_{th} = \frac{2.25}{kg_s\mathcal{L}^2(1+\gamma^2)}\left\{\frac{15L}{ct} + 2\alpha\ell + \ln\frac{1}{\sqrt{R}} + \ln 4\right\}^2$$ (7)

and $\gamma \approx 1$ (at threshold) is the ratio of backward to forward pump amplitude in the OPO cavity.

2.3 Phase Matching in KTP

Like most non-linear processes, optical parametric oscillation must obey phase matching constraints[5,6]. In the absence of dispersion, phase matching is automatically satisfied for collinear propagation of the three fields. Dispersion in the crystal causes non-uniform changes in the wavevectors of the pump, signal, and idler. Because of this non-uniform change, signal and idler wavevectors do not sum to equal the pump wavevector, even though their energies do. This results in a phase mismatch between the pump and shifted frequencies, which limits the interaction between them. Phase matching is satisfied when the refractive indices n_p, n_s, and n_i, of the pump, signal, and idler, respectively, satisfy

$$n_p\omega_p = n_s\omega_s + n_i\omega_i \qquad (8)$$

Since the refractive index in crystals is angularly dependent, the constraints of phase matching typically require the OPO crystal to be cut at a specific angle determined by dispersion; efficient frequency conversion can only be achieved within a narrow band of angles around this so called "phase-matched" angle. Furthermore, once this angle is determined phase matching can only be achieved for a quite narrow band of pump frequencies. However, in optical parametric oscillation the frequencies of the signal and idler are free to vary (as long as their sum is constant) so that phase matching can be obtained over a wide range of angles. Indeed optical parametric oscillators (OPO's) can be tuned in frequency simply by rotating the non-linear crystal. Thus OPO's provide an easily tunable, source of "new" frequencies.

Potassium Titanyl Phosphate (KTP) belongs to the orthorombic crystal system, and therefore is optically biaxial. The mutually orthogonal principal axes of the index ellipsoid are defined such that $n_x < n_y < n_z$, and the optical axes lie in the x-z plane[6] (at 18° to the z axis in KTP). The indices of refraction for any propagation direction are given by the index ellipsoid defined by:

$$k_x^2/(n_{\omega j}^{-2} - n_{x,\omega j}^{-2}) + k_y^2/(n_{\omega j}^{-2} - n_{y,\omega j}^{-2})$$

$$+ k_z^2/(n_{\omega j}^{-2} - n_{z,\omega j}^{-2}) = 0, \; j=p,s,i \qquad (9)$$

where $k_x = \sin\theta \cos\phi$, $ky = \sin\theta \sin\phi$, $k_z = \cos\theta$; θ is the angle of propagation with respect to the z axis and ϕ is the angle to the x axis in the x-y plane. The subscript "j" refers to either the pump, signal, or idler frequency. The equations 8 and 9 must be solved to determine the refractive indices ($n_{\omega j1}$, $n_{\omega j2}$) for the two eigen-polarizations perpendicular to the propagation direction for each wavelength, and this can be done numerically using the technique of Yao and Fahlen[7]. The indices of refraction are obtained from the appropriate Selmeier equations[8].

The phase matching condition for optical parametric conversion in KTP is given by:

$$\omega_p n_{\omega p1} = \omega_s n_{\omega s1} + \omega_i n_{\omega i2} \text{ (Type IIa)} \qquad (10)$$

or,

$$\omega_p n_{\omega p1} = \omega_s n_{\omega s2} + \omega_i n_{\omega i1} \text{ (Type IIb).} \qquad (11)$$

Type I interactions have very low non-linear coefficients in KTP. For propagation in the x-z plane (ϕ = 0) in KTP, Type II interactions correspond to Type II phase matching in a positive uniaxial crystal (e.g. $\omega_p n_p^o = \omega_s n_s^o + \omega_i n_i^e$), where "o" and "e" represent the ordinary and extraordinary rays respectively.

Figure 1 shows the calculated signal and idler wavelengths as a function of phase-matching angle θ for ϕ = 0 in a type-II KTP OPO pumped at 1.064 µm, with d_{eff} plotted in arbitrary units. The polarizations of the pump wave is along the y axis (o-wave) of the crystal as is that of the signal wave. The idler wave is polarized in the x-z plane (e-wave). The degenerate point at 2.128 µm corresponds to a phase matching angle of θ=54°. From the curves we see that the tuning range for the KTP OPO covers the transparency range of KTP to beyond 4.2 µm.

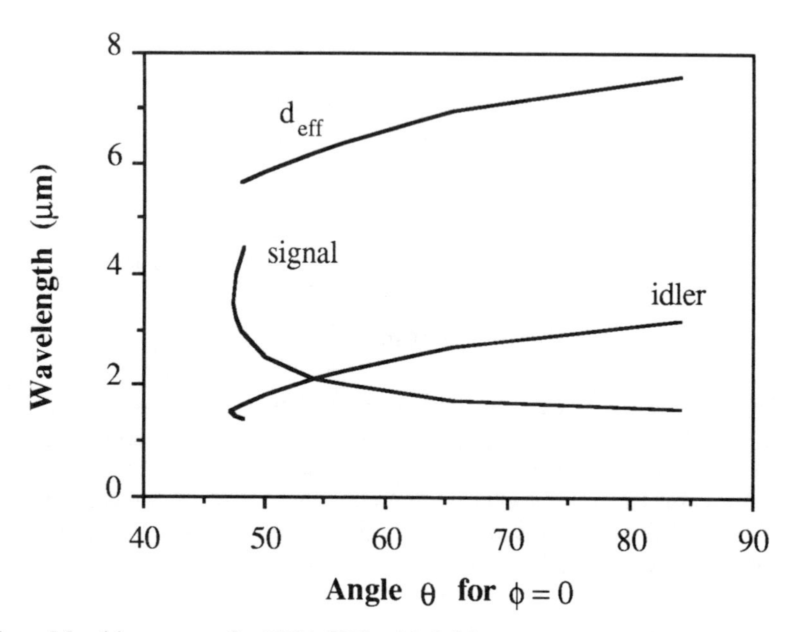

Figure 1 : Phase Matching curves for KTP OPO with 1.06 µm pump; d_{eff} is plotted in arbitrary units.

(3) EXPERIMENTAL

3.1 Pump Source : Diode Pumped Nd:YAG Laser

The pump laser is shown schematically in figure 2, and consisted of a 3.5 mm diameter Nd:YAG rod, side pumped by two 5 bar stacked, quasi-CW diode arrays. The laser cavity was hemispherical, using a 50 cm concave high reflector and 20 % transmitting flat output- coupler, separated by 50 cm. In order to obtain Q-switched operation, the cavity also contained a thin film polariser and a KD^*P- based electrooptic Q-switch. The maximum output energy of this device was 10 mJ, produced in a 50 ns pulse. The wallplug efficiency of this non-optimised device was rather low, approximately 3.2 %. However, wallplug efficiencies approaching 10 % can be achieved after optimisation.

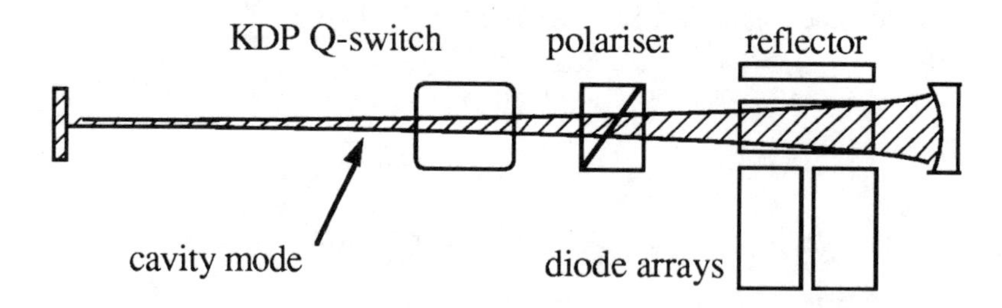

Figure 2 : Diode-Pumped, Q-switched Nd:YAG laser, for pumping OPOs.

3.2 Optical Parametric Oscillator : KTP

The OPO used in these experiments employed a 15 mm length KTP crystal as the non-linear element. KTP exhibits high damage threshold (up to the $GWcm^{-2}$ level), and has a relatively wide acceptance angle. For this application we were interested in producing an eyesafe source at 1.61 µm for laser radar. Thus tunability of the device was not an issue, indeed we required stable operation at the desired wavelength. To this end, the crystal was cut so as to achieve type II non-critical phase matching for a pump wavelength of 1.06 µm (Nd:YAG) and a signal wavelength of 1.61 µm; the idler wavelength was therefore 3.1 µm. This corresponds to operation in the wings of the OPO tuning curve, as shown in figure 1, making the output wavelengths quite insensitive to rotation of the crystal.

In practice the crystal could be rotated by 3 degrees with no measurable change in the output wavelength (angstrom resolution). In this non-critically phase-matched configuration the refractive indices are $n_s = 1.73$, $n_p = 1.73$, and $n_i = 1.82$. The effective non-linear coefficient is[8] $d_{eff} = 7.6 \times 10^{-12}\ mV^{-1}$. The round trip loss of the OPO crystal is 2al = 0.01, where we have included losses due to the optical coatings on the crystal faces. The mode-coupling coefficient g_s is approximately unity at threshold. In the present case of non-critical phase matching, the walkoff is negligible so that for small crystal lengths $\mathcal{L} \approx \ell$. Using these parameters, we calculate (using equation 7) a threshold pump intensity of $I_{th} = 20\ MWcm^{-2}$, for our OPO.

3.3 Plane-Parallel Cavity

The crystal was centred inside an optical cavity formed by a pair of plane-parallel mirrors, as shown in figure 3a. The input mirror, through which the pump enters, was anti-reflection coated at the pump wavelength (1.06 μm) and highly reflecting at the signal wavelength (1.6 μm); while the output mirror was highly reflecting at the pump wavelength and 10 % transmitting at the signal wavelength. The laser output was focussed by a 100 cm focal length lens, carefully positioned so as to mode match the waist of the pump to the cavity mode of the OPO resonator. This matching proved critical to successful operation of the OPO, and can be explained as follows: if the waist of the pump does not match the cavity mode then the pump will move out of phase with the signal as they propagate in the cavity, resulting in reduced interaction and lower conversion efficiency. For the plane-parallel resonator this matching required that the confocal region of the focused pump coincide with the mirror spacing of the resonator (see figure 3a).

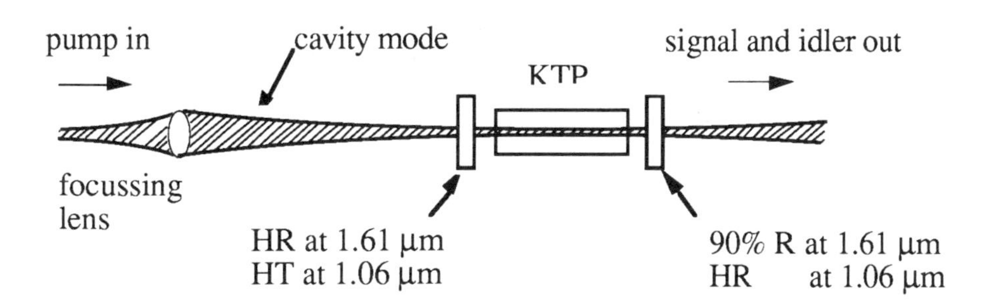

Figure 3a : Plane-parallel resonator for optical parametric oscillator, showing mode matching of focussed pump to OPO cavity mode.

3.4 Confocal Cavity

For this experiment the OPO cavity was changed to a confocal configuration by employing a pair of 5 cm concave mirrors with identical coatings to those used in the plane parallel cavity. The use of this cavity allows the pump to be more tightly focused while still maintaining good matching of the pump waist to the OPO cavity mode. This configuration is shown in figure 3b, and allows for tighter focussing of the pump, to reduce the pump energy threshold for the device.

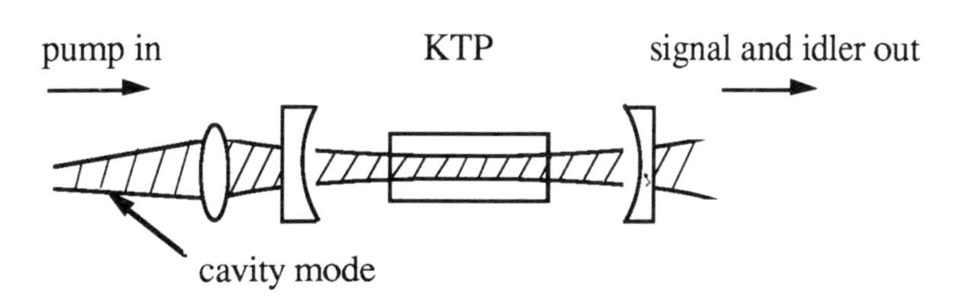

Figure 3b : Confocal OPO resonator allows tighter focussing of pump, while still maintaining good mode matching.

3.5 Intra-Cavity OPO's

In this case the OPO cavity forms part of the pump laser cavity, and the pump makes many passes through the OPO crystal. The cavity configuration employed in these experiments is shown in figure 4. The OPO resonator consists of a plane mirror (1), anti-reflection coated at the pump wavelength (1.06 μm) and highly reflecting at the signal wavelength (1.6 μm), and a 5 cm radius of curvature concave output coupler (2), highly reflecting at the pump wavelength and 90 % transmitting at the signal wavelength. The pump laser cavity is formed by mirror (2) and a third mirror (3) which is a 50 cm radius of curvature high reflector at the pump wavelength. The remainder of the cavity contains the conventional elements of a Q-switched oscillator; a Pockels cell and polariser.

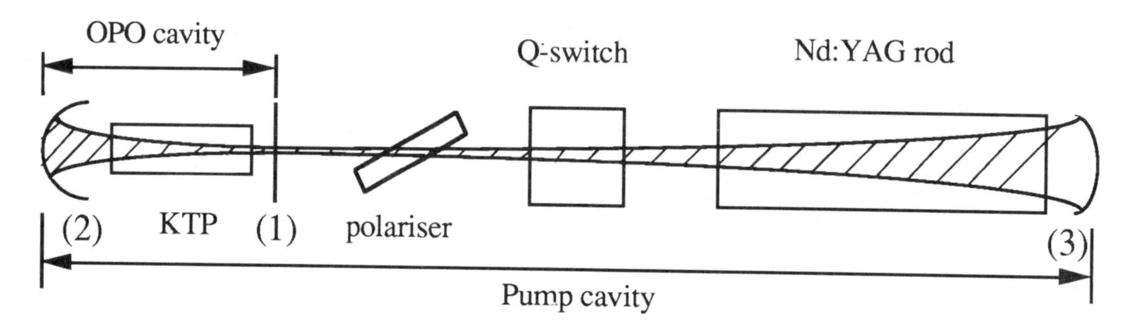

Figure 4 : Intra-cavity OPO, in which the OPO resonator forms an integral part of the pump-laser cavity.

4. RESULTS

4.1 Extra-Cavity OPOs

An oscillogram showing the input and depleted pump together with the generated signal pulse, is presented in figure 5. The pump depletion was approximately 16 % on the first pass and reached 40 % after the second pass, with the residual energy (15 %) going into the idler output at 3.1 μm. This shows the advantage of reflecting the pump back through the OPO crystal to improve conversion efficiency.The performance of the plane-parallel OPO cavity is described by figure 6, showing the output energy at 1.61 μm as a function of pump input energy at 1.06 μm. The OPO reached threshold at a pump energy of 1.5 mJ, with a maximum output energy of 2.5 mJ obtained at 10 mJ pump energy. This corresponds to an energy conversion efficiency of 25 %. Since the 25 ns duration signal pulse is only half as long as the pump pulse, the power conversion efficiency of the OPO is 50 %.

High energy, short pulse operation is desirable for laser radar, to meet range and spatial resolution requirements, respectively. While synch-pumped OPOs[9] have produced very-short, low-energy pulses (ps timescales) using mode-locked pumps, we are interested in pulses of order 1-5 ns with milli-joule pulse energies. In order to shorten the pulse length of the 1.61 μm source, we reduced the laser cavity length to 25 cm, maintaining a hemispherical cavity by reducing the curvature of the high reflector to 25 cm. This configuration produced a pump pulse of 18 ns FWHM, and generated a 13 ns 1.6 μm pulse when used to pump the OPO. This pulse shortening effect occurs because a large number of passes must be made in the OPO cavity, before the threshold for parametric oscillation is reached. In the oscillograms of OPO output shown in figure 5, that the peak of the signal pulse is delayed by several nanoseconds from the peak of the pump pulse. The slow buildup time reduces the pulselength of the signal and idler outputs, but also increases the threshold for short-pulse pumping (see $15L/c\tau$ term in equation 7). Unfortunately OPO's employing external cavities are characterised by fairly long buildup times, typically several ns, so that the OPO threshold increases for pump pulses shorter than about 10 ns. Furthermore, it is difficult to generate pump pulses shorter than about 10 ns using conventional Q-switched solid-state lasers. These considerations limit extra-cavity OPO pulses to durations greater than about 5 ns. In order to obtain shorter pulses from the OPO, and also to minimise threshold for quasi-CW operation of the OPO, we investigated intra-cavity parametric oscillation, as discussed in the next section.

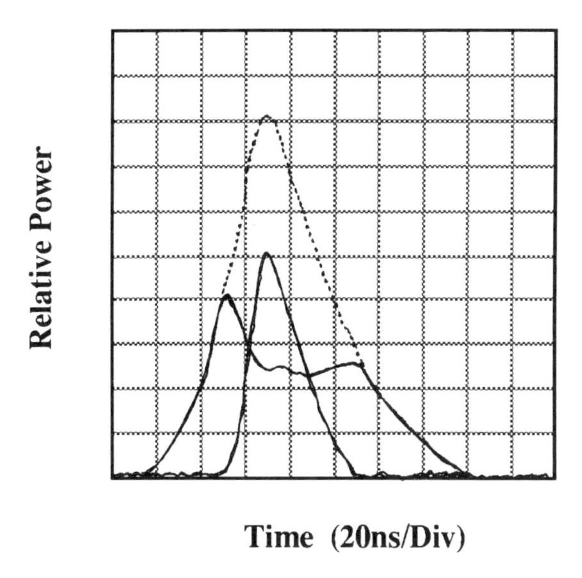

Figure 5 : Oscilloscope trace showing input and depleted pump and generated signal pulses for the plane-parallel OPO cavity.

The output energy at 1.61 μm of this device is plotted in figure 6, together with that obtained from the plane-parallel cavity, as a function of input pump energy. The threshold pump energy of this device was only 0.8 mJ, considerably lower than that of the plane-parallel configuration. A maximum output energy at 1.6 μm of 1.8 mJ was obtained for 5.5 input pump energy. This corresponds to an energy-conversion efficiency of 35 %.

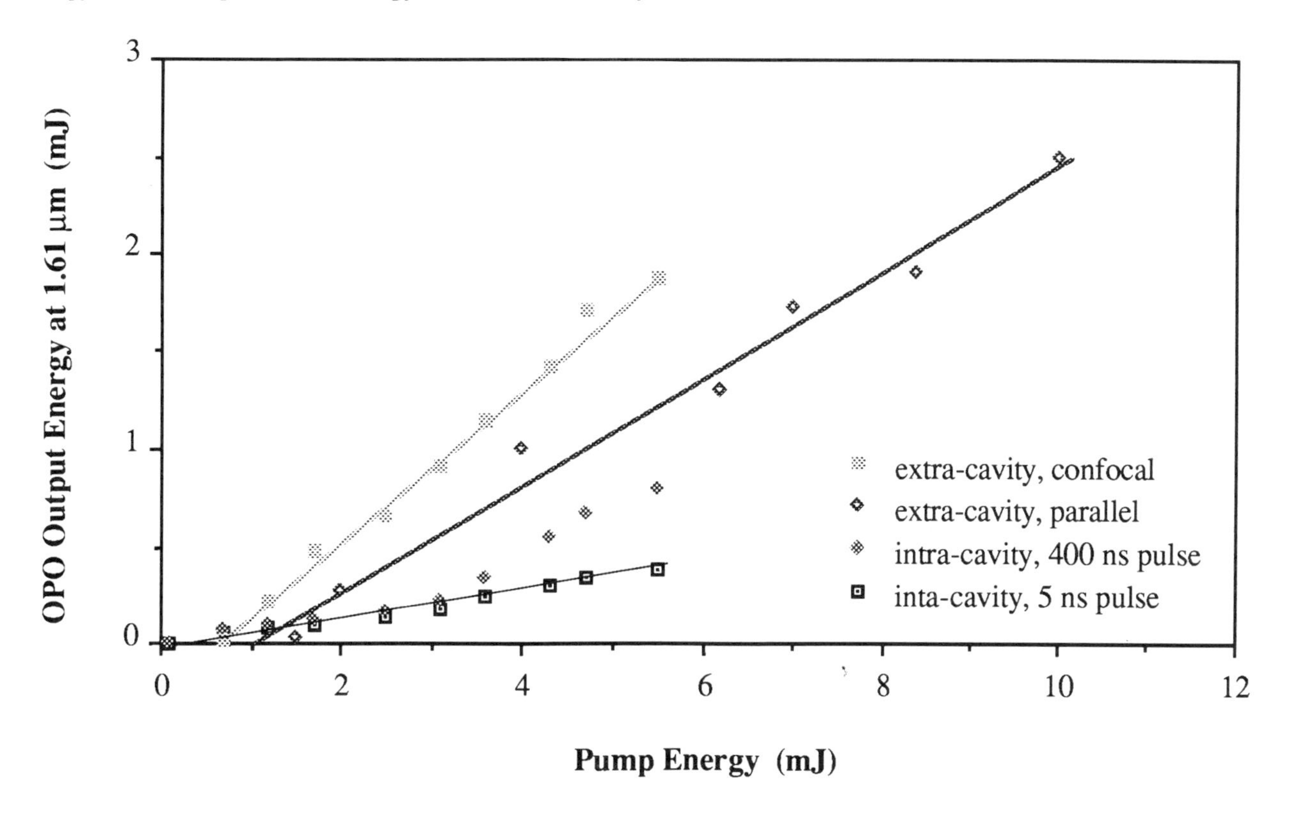

Figure 6 : Signal (1.6 μm) energy vs. pump energy for various OPO configurations, as marked.

The wavelength of the signal was measured to be 1.61 ± 0.01 μm, using a 0.25 m spectrometer operating in the 1 to 2 μm band. From this signal wavelength, we calculate an idler wavelength of 3.14 μm. While the OPO output is principally at the signal and idler frequencies, there are a number of other non-linear frequency-mixing effects occurring in the KTP crystal. We measured weak output at 0.532 μm, corresponding to the second harmonic of the 1.064 μm pump, and also at 0.804 , and 0.640 μm. These latter wavelengths could be explained by the presence of 1.608 μm radiation which is doubled to produce 0.804 μm output, and mixes with the 1.064 μm pump to sum-frequency generate 0.640 μm output. These other non-linear processes are far from phase matched, thus the energy lost to them is minimal.

4.5 Intra-Cavity OPOs

The dynamics of this laser system appear to be quite complex, however, the system appears to operate in three basic regimes, determined by the conversion efficiency and threshold of the OPO, which in turn control the intra-cavity pump power. The dependence of 1.6 μm pulselength upon pump power is plotted in figure 7, where the pump power was varied by changes in stored energy. The two regimes shown are for high and low OPO threshold, where the OPO threshold was reduced by moving the crystal out of the high-intensity beam waist. Initially, the pulselength increases with pump power, reaching 30 ns when the pulselength drops abruptly to 8 ns. Further increase in pump power causes reduction in pulselength, with 5 ns pulse durations achieved at the highest pump powers investigated.

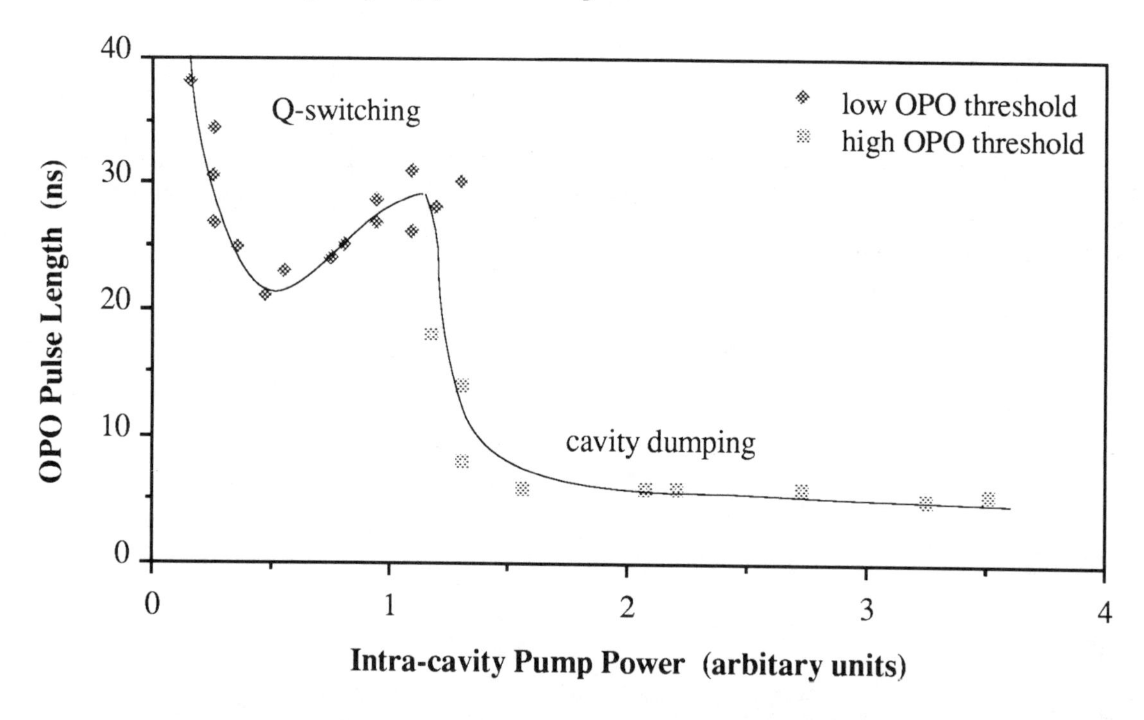

Figure 7 : Pulsewidth vs. intra-cavity flux for various OPO thresholds.

The performance of the OPO in each regime is illustrated by the oscilloscope traces shown in figures 8 a, b, and c, which show the signal and pump waveforms. The three operating regimes are :

Figure 8a : Intermediate pulse length, increasing with pump power.
Experimentally, operation in this regime is achieved by placing the OPO crystal at the waist of the pump. In this case, the threshold of the OPO is comparable to that of the pump. Increasing the pump power causes additional after-pulses to be emitted. We have obtained single pulse energies in the 0.2 to 0.3 mJ range, in this regime.

Figure 8b : Long pulse length, increasing with pump power.
In these experiments we increased the OPO threshold slightly, by moving the crystal away from the waist of the pump beam. Initially the pump exhibits relaxation oscillations over approximately 100 ns. These quickly die out and the pump

intensity approaches the steady-state. The OPO output reflects the trends of the pump, and consists of a pulse train followed by a long pulse of some 300 ns duration. Pulselengths of up to 500 ns have been achieved using this technique.

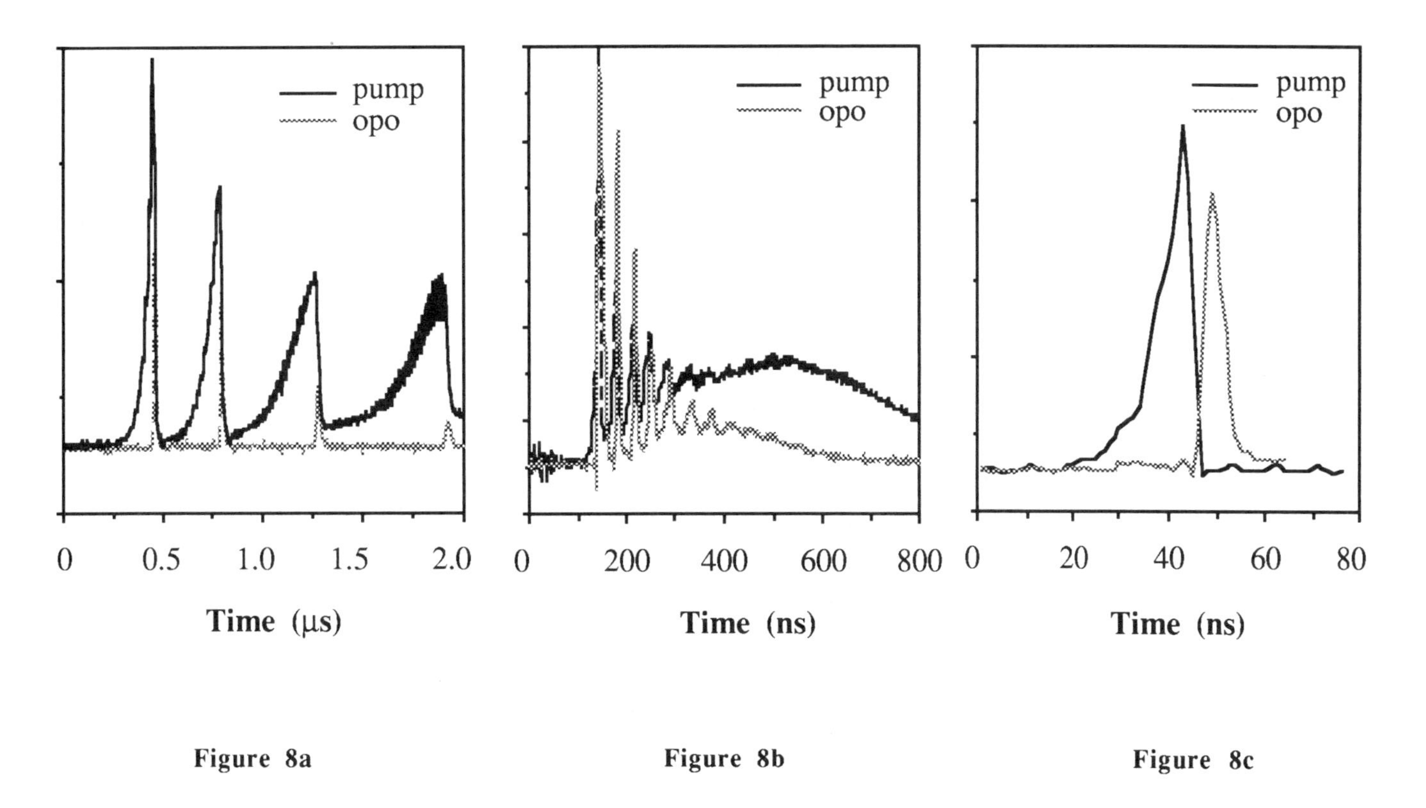

Figure 8a Figure 8b Figure 8c

Figure 8c : Short Pulselength, decreasing with pump power

In these experiments, we again increased the threshold for the OPO by moving the crystal further away from the waist of the pump beam. This allows true Q-switched operation of the OPO by reducing the overall pump depletion. The pulse energies obtained for short pulse (7 ns) and long pulse (500 ns) operation of the intra-cavity OPO, are plotted in figure 6, together with the data obtained for the extra-cavity OPOs discussed previously.

5. DISCUSSION

5.1 Extra-Cavity OPOs

Higher conversion efficiencies were observed with the confocal resonator than with the plane-parallel resonator, due to the lower threshold energy required in the former configuration. Threshold energy is reduced because the pump is focussed more tightly into the KTP crystal to obtain optimum matching for the confocal resonator. In the confocal configuration, the power conversion efficiency of 1.06 μm pump to 1.61 μm signal exceeds 50 %, and the overall power conversion efficiency of the OPO, to both signal and idler outputs, exceeds 70 %. Since the wallplug efficiency of the laser is 3.2 %, the overall wallplug efficiency of this 1.61 μm source is 1.1 % which is a benchmark for this type of device. Energy conversion efficiencies of 25 %, and peak-power conversion efficiencies of 50 % were observed for the 1.6 μm signal produced in the plane-parallel OPO resonator.

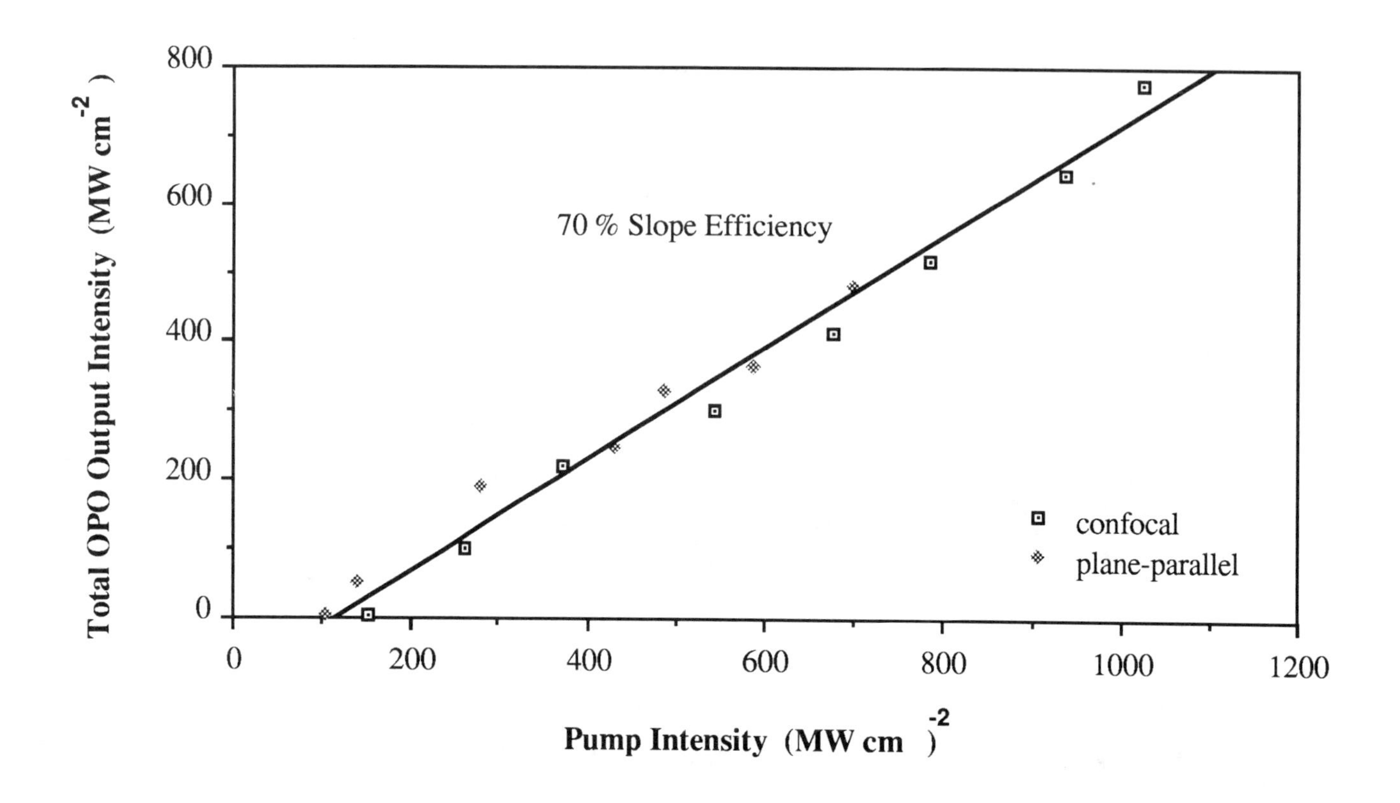

Figure 9 : Total OPO intensity at both signal and idler wavelengths vs. pump intensity.

Optical parametric oscillation is intensity dependent, so pump intensity is a more meaningful parameter for gauging OPO performance than is pump energy. The influence of pump intensity upon total OPO output intensity is plotted in figure 9. The intensities were determined from measurements of the focal spot-size (200 ± 40 μm) measured using a scanning pinhole. The OPO threshold is exceeded for pump intensities around 100 ± 40 $MWcm^{-2}$, and the total output power increases with a 70 % slope efficiency. This threshold pump intensity is a factor of 5 greater than that calculated from equation (7). Recent measurements by Eckhardt[10] suggest that the effective non-linear coefficient of KTP is actually $d_{eff} = 3.1 \times 10^{-12}\ mV^{-1}$, a factor of 2.4 lower than that used to calculate our pump intensity threshold. Using Eckhardt's measured coefficient, we obtain a threshold of 115 $MWcm^{-2}$; in much better agreement with our observations. These results support Eckardt's claim[10] that the non-linear coefficient of our KTP sample is lower than the generally accepted value.

5.2 Dynamics of Intra-Cavity OPOs

A complete description of the behavior of the intra-cavity OPO requires analysis of the transient nature of the coupled-wave equations that describe parametric oscillation. In this brief discussion we will simply describe our interpretation of the behavior observed, and defer detailed calculation for a later paper.

5.2.1 Low OPO Threshold : Intermediate Pulselength

In the absence of pump depletion due to the OPO, the pump power obtained in Q-switched mode grows exponentially with time. When the OPO threshold is low, the pump power cannot reach the levels that are typically achieved in typical Q-switched operation, because the OPO depletes the pump. The pump is limited to the low power regime, where the buildup of pump power is relatively slow; figure 8a shows a decreasing rate of pump growth for each successive relaxation oscillation. As noted above, the OPO requires a certain number of passes of the pump in order to build up gain. This buildup time allows the pump to exceed the OPO threshold before any significant pump depletion occurs. However, the onset of parametric oscillation soon begins to deplete the pump as 1.6 μm light is coupled out of the cavity. This increased intra-cavity loss rapidly terminates the pump. The slow buildup of pump power and gradual pump depletion produce an OPO output of intermediate pulse length, typically 50 ns (see figure 7). As pump power is increased, the parametric gain becomes higher, reducing the buildup time of the OPO. The OPO depletes the pump more rapidly, reducing the pulselength of the OPO output to a minimum of about 20 ns. Once this minimum is reached, further increases in pump intensity begin to lengthen the OPO pulse, because the increase in pump power also increases the rate of growth of pump intensity, so that a higher pump power is reached during the buildup time of the OPO. The OPO takes longer to deplete this greater pump power, thus the OPO output *increases* in pulselength at higher pump powers, as shown in figure 7.

The majority of the laser energy remains stored in the medium because the pump is terminated before full extraction can be achieved. The pump power begins to build up again until a second OPO output pulse is produced. This process continues until all of the stored energy is extracted, and a train of intermediate- pulselength outputs are produced. Increasing the pump power results in a marginal increase in the peak power of the OPO pulses, but most of this additional energy goes into producing more after-pulses.

5.2.2 Intermediate Threshold : Long Pulse and Variable Pulselength Operation

In the previous case the threshold of the OPO was comparable with the laser threshold, thus the onset of parametric oscillation terminated the pump. If the pump depletion is reduced, or occurs more slowly, then the laser pulse will not be truncated. Rather, the intensity of the pump will be clamped near the threshold intensity of the OPO. This results in an increase in the pump and OPO pulse lengths, as the stored energy must still be dissipated as flux in the laser cavity. Thus the intra-cavity OPO also provides variable-length pulses. The OPO output consists of a series of relaxation oscillations, damping out as the pump reaches a steady-state level at the threshold of the OPO, and the remaining pump energy is extracted in a single long pulse.

5.3.3 High OPO Threshold : Short Pulse Operation

At low pump powers the rate of increase in pump intensity is slow, and the buildup time of the OPO is long. As the pump exceeds threshold for the OPO, there is a gradual buildup of 1.61 μm power as the pump is depleted. This slow rate of pump depletion produces an intermediate OPO pulselength (see 5.2.1, above) before the pump is truncated. As the pump power is increased in this low power regime, the pump reaches a higher level before depletion begins, but the buildup time of the OPO remains fairly long. Thus the OPO output pulse lengthens because more pump energy must be extracted before the pump is terminated. At higher pump powers true Q-switched operation is achieved, and the laser intensity grows so rapidly that a far greater pump power is reached before depletion sets in. Due to this greater pump intensity the OPO buildup time is dramatically reduced so that depletion and subsequent truncation of the pump follows rapidly once threshold is exceeded. In this mode, the OPO effectively cavity-dumps the laser output. The device therefore produces very short pulses at 1.06 and 1.61 μm. The duration of a Q-switched pulse decreases with increasing gain so that the pulselength of the OPO output decreases at higher pump powers, as shown in figure 7.

The output coupling of this device was far from optimum, and use of a lower reflectivity output-coupler would result in higher extraction energies, together with shorter pulselengths. The onset of relaxation oscillations discussed above, could be forestalled in this manner.

6. SUMMARY and CONCLUSIONS

In this series of experiments we have observed total peak-power conversion efficiencies (to signal and idler) exceeding 70 %, and wallplug efficiencies greater than 1 %. The wallplug efficiency is currently limited by the 3.2 % wallplug efficiency of our non-optimised pump laser. Laser efficiencies of up to 10 % are possible with optimisation, so that wall plug efficiencies approaching 5 % are well within the reach of the OPO. We have also shown that the intra-cavity OPO is a versatile device, capable of producing short pulses (5 ns) at high efficiencies The present system is far from optimised, so that efficiencies approaching those of the extra-cavity OPO should be possible. Current efficiencies stand at a respectable 32 %. Certainly, the pump energy threshold of this device is considerably lower than the extra-cavity OPO, which makes it ideal for high repetition-rate operation. We have shown that short pulse operation (<5 ns) can be best achieved using intra-cavity OPO's. We have also demonstrated the intra-cavity OPO's potential for producing variable pulselength output, with pulselengths ranging from 5 ns to 500 ns being obtained. These devices also exhibit lower thresholds than their extra-cavity counterparts, and can therefore be used to achieve high repetition rate operation. We have demonstrated pulse train generation with interpulse separations of order 20 ns, and little degradation in energy over several pulses. This suggests that operation at repetition rates as high as 10-20 kHz should be possible with little reduction in pulse energy.

Intra-cavity OPO's are clearly a rich area of technology, and will attract major attention in the near future. Given the need for tunable sources in the infra-red, together with pulse length agility and narrow line operation, these devices are well suited to laser countermeasure applications. In this configuration the pump makes many passes through the OPO, causing a significant reduction in the OPO threshold. Since the pump cavity consists of two total reflectors the laser threshold is also extremely low. As a result of these two factors, this device is ideally suited to high repetition rate applications where the peak power of the pump is reduced. In such a system the laser would be CW pumped by a diode array, and Q-switched at the desired repetition rate. Thus the intra-cavity OPO presents the possibility of short pulse, high repetition rate, tunable output.

OPOs provide output energies at the millijoule level, with shot-to-shot amplitude stability of $\pm$ 5 %. We believe these high efficiencies are due primarily to the large non-linear coefficient of KTP when cut for Type II non-critical phase matching, and to the excellent beam quality obtained from the diode-pumped laser. With the current interest in efficient, lightweight, rugged, eyesafe laser sources, the all-solid-state approach of a diode-pumped Nd:YAG laser and KTP optical parametric oscillator, is quite attractive. The high efficiencies and output energies, together with the tunability of OPOs, bring a large number of new applications within the reach of diode-pumped Nd:YAG lasers.

7. ACKNOWLEDGEMENTS

This work was funded entirely by Fibertek, Inc., internal research and development. The authors gratefully acknowledge the support and enthusiasm of our technical staff : Khoa Le, Elias Fakhoury, George Babec, Anh Tran; and Mindy Levinson for her proofreading.

8. REFERENCES

1. R.C. Stoneman and L. Esterowitz, in *Proceedings of the Conference on Advanced Solid State Lasers*, Optical Society of America, Washington D.C. (1990).

2. Z. Chu, U.N. Singh, and T.D. Wilkerson, Opt. Commun., 75, pp. 173-177 (1990).

3. A. Yariv, "Quantum Electronics", Second Edition, John Wiley & Sons Inc., New York (1975).

4. S. J. Brosnan and R. L. Byer, IEEE Journal of Quant. Electr. QE-015, pp. 415-443 (1979).

5. R. L. Byer in, "Quantum Electronics, Volume I." edited by Herbert Rabin and C. L. Tang, Academic Press, New York (1975).

6. Max Born and Emil Wolf, "Principals of Optics", Macmillan Co, New York (1964).

7. J. Q. Yao and T. S. Fahlen, J. Appl. Phys. 55, pp. 65-72 (1984).

8. K. Kato, IEEE Journal of Quant. Electr, QE-24, pp. 3-10 (1988).

9. G.T. Maker, and A.I. Ferguson, Appl. Phys. Lett., 56, pp. 1614-1616 (1990).

10. R.C. Eckhardt, H. Masuda, Y.X. Fan, and R.L. Byer, IEEE Jnl. Quant. Electron., QE-26, pp. 922-933 (1990).

A HIGH SPEED SHORT RANGE LASER RANGE FINDER

Robert M.A.M. Gielen,
Ronald P. Slegtenhorst

Delft Instruments Electro-Optics BV
Van Miereveltlaan 9, 2612 XE Delft, The Netherlands

ABSTRACT

A high precision multi-target, short-range laser range-finder using a semiconductor laser diode is described. The laser diode is intensity modulated with a time-dependent frequency voltage signal. The return bundle is detected by a semiconductor photo diode and mixed with an undelayed fraction of the time-dependent frequency signal. This produces sum and difference frequencies, of which the difference frequencies are filtered out and are analyzed for individual components by a fast fourier transform processor. Each individual frequency component represents a specific distance to a target. The optical transmit and receive bundles are coaxial and can be scanned by a mirror scanner up to 100 Hz in elevation and 10 Hz in azimuth over a 30° by 30° field of view.
The estimated accuracy in distance is 10 cm with a 1 ms 1.5 Ghz chirp and 36 mW of optical power at a wavelength of 1310 nm, over distances ranging from 1 to 50 m. The fast fourier transform processor estimates the up to 512 individual frequency components in less than 1 ms.
The non-linear time-dependent frequency behavior of the voltage controlled oscillator is compensated by an optical feedback path. This contributes greatly to the accuracy of measured distances.
This study has been carried out in the framework of IEPG activities on autonomous guided vehicles.

2. INTRODUCTION

Short range Laser Range Finders (LRF) are currently being developed for use in Autonomous Guided Vehicles (AGV). They present a 3D picture of the space in front of the AGV. Current LRF are merely based on OTDR, using short optical pulses. For the short ranges applicable to the use of AGV's, very short optical pulses are needed. Very often in the sub nano second range. This severely limits the total amount of energy in individual pulses. This is certainly the case in semiconductor laser diodes. Integration of several distance measurements is possible, but severely limits the measurement rate and dramatically increases the amount of optics and electronics. The fundamental limitation is the energy-bandwidth product of semiconductor laser based OTDR. The use of puls compression techniques circumvents this limitation. Radar pulse compression techniques are used to realize the large energy-bandwidth product needed[12,13]. The techniques of Frequency Modulated Continuous Wave (FM-CW) Radars[1,2,5] are transferred to an Intensity Modulated (IM) semiconductor laser diode. Thus increasing the received optical energy from targets by several orders of magnitude and simultaneous decreasing the measurement time.

To get real-time 3D pictures fast scanners are also needed. Conventional scanning techniques use rotating (polygons) or oscillating mirrors, at least for one axis[9,11]. This technique makes it nearly impossible to change the scan pattern or frequency in the case of an oscillating mirror, once the scanner parameters are chosen. An alternative pattern can be highly desirable in an AGV.

3. DESCRIPTION

Figure 1 depicts a possible architecture for a LRF based on a FM-CW IM laser diode. A semicon-ductor laser diode is IM by a chirped VCO. The optical signal of the laser passes trough suitable optics and is aimed via a scanner at the target. The return beam which is composed of the reflections of multiple targets travels via the same path back towards the optical receiver (Avalanche Photo Diode, APD) where the optical signal is rectified to an electrical RF signal. This RF signal is the sum of multiple delayed and overlapping replicas of the original chirped VCO output signal.
Because the round-trip delay is much less (max 1 microsec) than the chirp duration (about 1 millisec), it is possible to mix the chirped VCO signal with the received sum of chirps. This produces a series of low frequency tones, whose frequencies are proportional to the distances of the targets. A target at a distance R produces a tone of frequency f, which is ruled by:

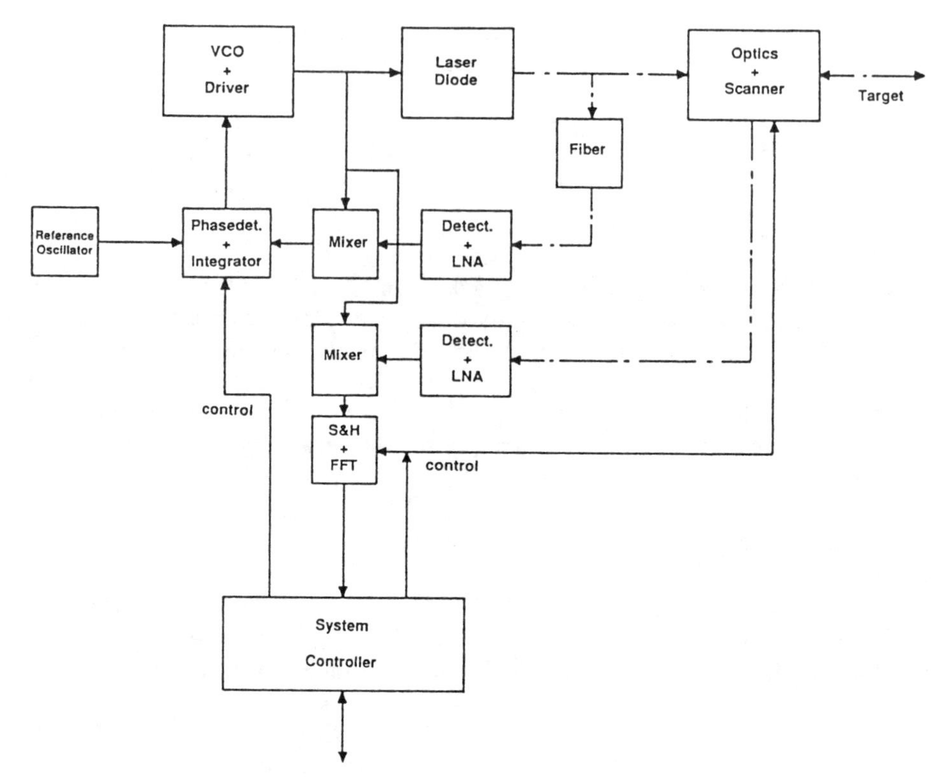

Figure 1 Architecture of the Laser Range Finder

$$f = \frac{2Rb}{c'} \tag{1}$$

In (1) b is the frequency gradient of the chirp, c' the velocity of the laser light in the medium between laser and target.

Before the output of the APD is send to the micro-wave mixer it is first amplified by the Low Noise Amplifier (LNA).
The output of the mixer is lowpass filtered and send to the Fast Fourier Transform (FFT) processor. This FFT processor is composed of fast basic FFT building blocks[4,7]. First the signal is sampled and stored in memory then the samples are multiplied by a suitable window in order to reduce the effects of spectral leaking after FFT[2,3,8,10]. In the case of distinct frequency components wich do not overlap each other, it is relative easy to increase the resolution by means of interpolating the samples in the frequency domain, by making use of the FFT of the window function[8]. Figure 2 shows a typical spectrum at the output of the lowpass filter, generated by a single chirp and processed by a Hamming window. There are 6 individual distances in figure 2: at 1.1 m, 4.9 m, 5.123 m, 30.1 m, 31.1 m and 50.1 m. The responses to 4.9 and 5.123 m do overlap each other. 30.1 and 31.1m are clearly seperated. The S/N is 20 dB at the output of the lowpass filter.

The feedback path to linearize the frequency chirp in the time-domain is composed of a fiber, a photo detector, LNA, mixer, phase detector and reference oscillator. The idea behind this feedback is that a linear chirp produces a constant tone out of the mixer when the distance, producing this tone is constant. The fiber presents the virtual constant distance (the effects of change of length and index of diffraction with temperature are neglected). The tone proportional to the fiber length is compared with a reference tone, produced by the reference oscillator, in the phase comparator. The phase comparator output is integrated and this integrated signal corrects the chirp's momentary frequency, producing a highly linear chirp with time.

The system controller is a general purpose microprocessorboard. It controls the main parts of this LRF. That is to say: initiali-

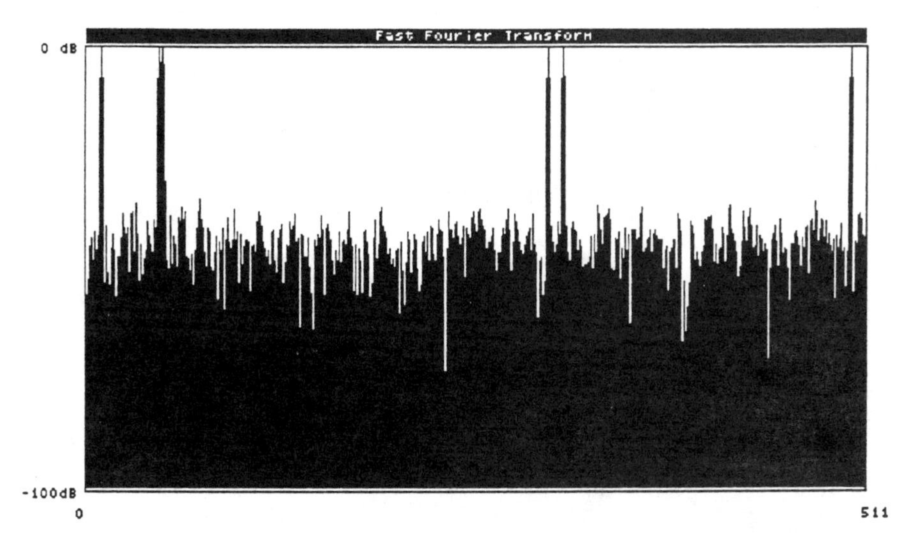

Figure 2 Simulated output of the FFT processor

zing the LRF, starting the chirp, starting the FFT processor and controlling the scanner. Further the system controller communicates with other systems in the AGV. For instance the distance interpretation and processing electronics. It receives commands of where to aim at, which kind of scan pattern etc.

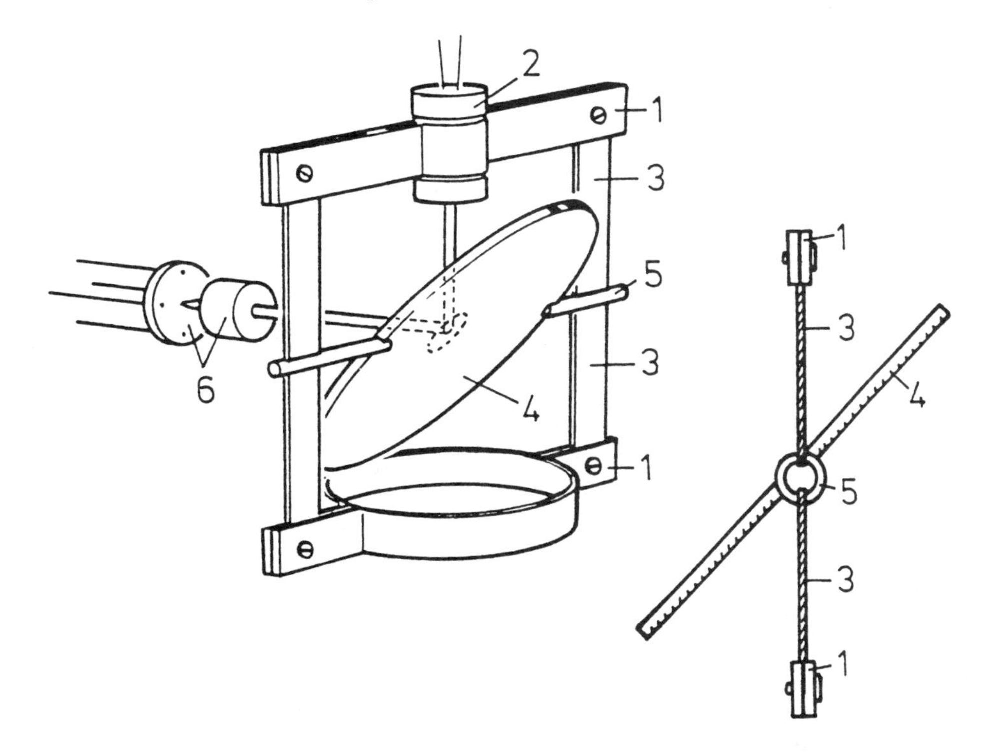

Figure 3 Basic idea of the scanner

The basic idea of the scanner is depicted in figure 3. The elevation driver is assembled from bending piezo actuators, in such a way that a mirror is rotated over a limited angle by these actuators. The advantage of using piezo actuators over for instance magnetic actuators is their low mass with respect to their driving torque. This low mass is of advantage in the orthogonal

direction, where it adds much less to the total inertia, especially because the actuators are located at the circumference of the mirror assembly. A low mass position read-out sensor is composed of an axial mounted collimator pen laser and an optical position detector at the rear of the scan mirror. This position sensor forms a position feedback to the piezo driver electronics.
In figure 3 are: 1) Azimuth frame, 2) Collimator pen Laser diode for position read out, 3) Piezo bending actuator, 4) Scan mirror, 5) Slotted hollow axis and 6) X-Y photo diode with optics for position read out.

4. PERFORMANCE ANALYSIS

4.1 Optical power budget and S/N

The expected optical power reflected by ideal diffusing targets and entering the receiver aperture is given by:

$$P_R=(\frac{D_o}{2R})^2 P_T \tag{2}$$

In (2) P_R=received optical power, P_T=radiated power from scanner, D_o=diameter of receiving aperture and R=distance of target.
P_R for D_o=3.2 cm and R=50 m is about 10^{-7} times the radiated power P_T.
The expected signal-to-noise ratio at the output of the LNA is given by[6]:

$$\frac{S}{N} = \frac{0.5*(\frac{qnmP_o}{h\nu}M)^2}{2q(I_d+\frac{qn(P_o+P_b)}{h\nu})F(M)M^2B+\frac{4kTBF}{R}} \tag{3}$$

Table 1 gives an overview of the symbols used in (3) together with expected values where applicable. The given values led to a S/N=12.6 dB. This S/N is defined as the ratio of the received signal power to the noise power in a band of width B. The band-width B is equivalent to a frequency resolution cell in the FFT.

4.2 Analysis of the VCO feedback

Figure 4 depicts the blockdiagram of the feedback path of the VCO linearizing circuit.
The momentary frequency of the VCO can described by:

$$\omega_m(t)=\omega_s t+f(u_i(t))t \tag{4}$$

In (4), $f(u_i)$ is a non-linear function of the VCO control voltage $u_i(t)$. $u_i(t)$ is the output of the integrator. Further ω_s is the start frequency of the frequency chirp and τ the round trip delay time. The delayed input signal to the mixer stage is than:

$$B\cos(-\omega_s(t-\tau)+f(u_i(t-\tau))(t-\tau)) \tag{5}$$

The input to the integrator is:

$$C(\omega_{ref}t+\omega_s\tau+[f(u_i(t))-f(u_i(t-\tau))]t+\tau f(u_i(t-\tau))) \tag{6}$$

In (6) ω_{ref} is the reference frequency to the phase comparator. (4), (5) and (6) lead to the delayed integral equation:

$$u_i(t)=C\int(\omega_{ref}t+\omega_s\tau-\frac{f(u_i(t-\tau))-f(u_i(t))}{\tau}\tau t+\tau f(u_i(t-\tau)))dt \tag{7}$$

To solve this non-linear integral equation, we carry out some approximations. In a practical system, the delay time τ tends to zero with respect to the measurement T_m. So we approximate
$\frac{f(u_i(t-\tau))-f(u_i(t))}{\tau}$ by $\frac{df(u_i(t))}{dt}$ and neglect $\tau f(u_i(t-\tau))$ and $\omega_s\tau$. Carrying out these approximations lead to the differential e-quation:

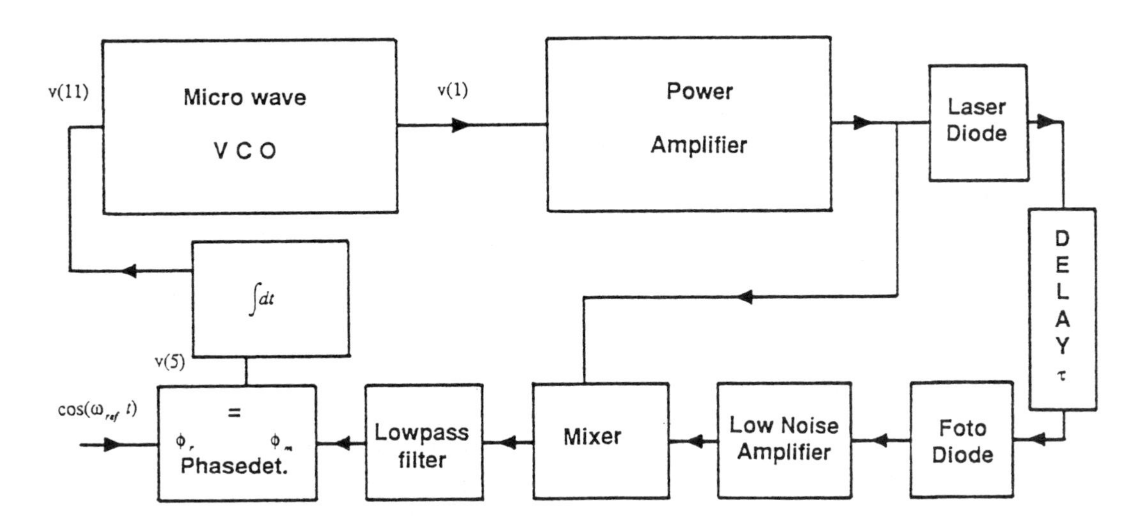

Figure 4 Blockdiagram of the feedback path of the VCO linearizing circuit

$$\frac{du_i(t)}{dt} = \frac{C\omega_{ref}t}{1+C\frac{df(u_i)}{du_i}\tau t} \tag{8}$$

$\frac{df(u_i)}{du_i}$ describes the non-linear behavior of the VCO. Since we are not interested in transients as long as they don't take to much time with respect to T_m, we only look at the behavior of $f(u_i)$ for $t \rightarrow\infty$.
We therefore approximate (8) by:

$$df(u_i) = \frac{\omega_{ref}}{\tau}dt \tag{9}$$

From (9) it is clear that

$$f(u_i(t)) = \frac{\omega_{ref}}{\tau}t \tag{10}$$

Figure 5 shows the results of a simulation of the feedback circuit. This figure clearly shows the linearizing effect of the feedback. The transient behavior is also acceptable.

In figure 5: v(1) is the output frequency of the VCO, v(11) the input voltage to the very non-linear VCO and v(5) the the output voltage of the phase detector. The non-linear VCO is described by: $y=x+x^2+x^3+x^4$.

5. SCANNER

Figure 3 shows the new scandrive idea. This new idea is based on the use of piezo electric actuators (bimorph bending types) as the driving element in the elevation direction. In the past actuators where of the electromagnetic type[11] (e.g. brushless DC motors).
The piezo actuators drive the mirror in elevation, by exerting a torque on the elevation shaft, via slots in this hollow shaft. The actuators are arranged in such a way that if the upper actuator exerts a torque CW by bending to the right, the lower actuator

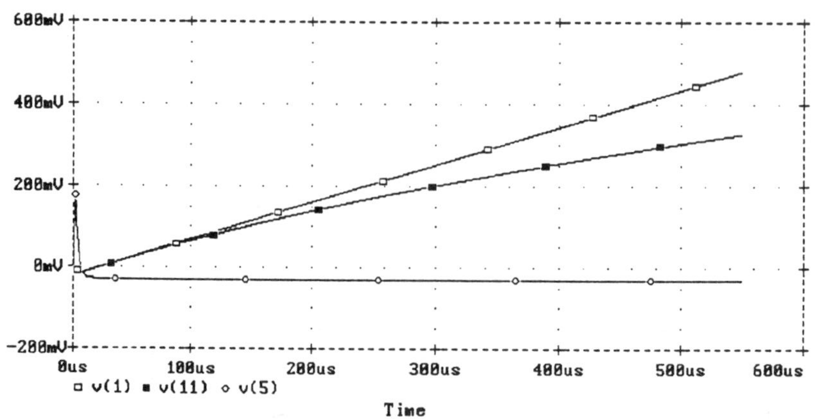

Figure 5 Simulation result of feedback loop

also exerts a torque CW by bending to the left. The diameter of the hollow shaft has to be small because the piezo actuators are limited in range (to a maximum of a few tenth's of a mm). But because they are capable of delivering a high torque, it is possible to drive the mirror this way with a considerable velocity and acceleration.

The relation between angular excursion of the mirror and applied voltage to the actuators suffers from non-linearities and hysteresis. We therefore incorporated a position read-out. This read-out is based on the deflection of a collimated laser bundle at the rear of the mirror. A small polished surface at the rear of the scan mirror acts as a plane mirror for the read-out device. The surface of this little mirror lays in the elevation axis. The laserpen collimator bundle is coaxial with the azimuth axis and is reflected by the mirror just at the crossing of the elevation and azimuth axis. This assures that there will be no translation of the reflected laserpen collimator bundle but only deflection. A X-Y position sensitive photo diode together with optics detects the reflected laser bundle. The X-Y position of the laser spot on the photo diode is a unique measure for the angular position of the main mirror.

6. EYE SAFETY

For the unaided eye, the LRF is eyesafe, even in a non-scanning mode. This is the case because the maximum energy over a 7 mm aperture doesn't exceed 19.7 mJ at the laser wavelength of 1310 nm in the maximum time of 10 s, according to the ANSI and FDA standards[14,15]. The value of 19.7 mJ is compatible with the optical power of 36 mW (see abstract) at our LRF output aperture of 3.2 cm.

7. CONCLUSION

We have described a concept of using laser diodes to measure distances for Autonomous Guided Vehicles. The implementation of this concept is capable of measuring the distance to several targets at a time, with high resolution.

A new idea is the way of driving a mirror in elevation and azimuth. With this new actuator idea it is possible to achieve high scanning rates, even for arbitrary scan patterns. Up to now high speeds were only possible with rotating mirrors (which can only used for fixed patterns).

The presented concepts are the result of calculations in a study, which has been carried out in the framework of IEPG activities on Autonomous Guided Vehicles.

8. ACKNOWLEDGEMENTS

We acknowledge financial support from NIVR and we thank Mr R.J. Lerou of FEL/TNO for the fruitful discussions.

9. REFERENCES

1 Berkowitz,R.S.; "Modern Radar"
2 Childers, D.G.; "Modern Spectrum Analysis", IEEE Press, 1978.
3 Kesler, S.B.; "Modern Spectrum Analysis,II", IEEE Press, 1986.
4 Plessey Semiconductors, "Digital Signal Processing IC Handbook", 1988
5 Skolnik,M.I.; "Introduction to Radar Systems"
6 Yariv,A.; "Optical Electronics", chapters 10 and 11
7 ZORAN Corporation, "Digital Signal Processors Data Book", 1987.
8 Andria,G.; Savino,M.; Trotta,A.; "Windows and Interpolation Algorithms to Improve Electrical Measurement Accuracy", IEEE trans. on Instr. and Meas., Vol.38, no.4, August 1989, pp.856-863.
9 Blais,F.; "Control of low inertia galvanometers for high precision laser scanning systems", Optical Engineering, vol.27, no.2, February 1988, pp. 104-110.
10 Harris,F.J.; "On the Use of Windows for Harmonic Analysis with the Discrete Fourier Transform", Proceedings of the IEEE, Vol. 66, january 1978, pp. 51-83.
11 Reich,S.; "The use of electro-mechanical mirror scanning devices", SPIE Vol.84 Laser Scanning Components & Techniques (1976), pp. 47-56.
12 Doyle McClure,J.;"Diode Laser Radar: Applications and Technology", SPIE Vol.1219, Laser-Diode Technology and Applications II (1990), pp. 446-456.
13 Abbas,G.L.;Randall Babbit,W.;Chapelle,M.de la;Fleshner,M.L.;Doyle McClure,J.;Vertatschitsch,E.J.; "High-precision fiber-optic position sensing using diode laser radar techniques", SPIE Vol.1219, Laser-Diode Technology and Applications II (1990), pp. 468-479.
14 ANSI z136.1 - 1986.
15 FDA, Federal Register Vol. 50, no. 161, Aug. 20, 1985.

Symbol	Definition	Value	Unit
q	charge of an electron	$1.6*10^{-19}$	C
n	quantum eff.	80	%
h	Planck's const	$6.63*10^{-34}$	Js
ν	freq. laserlight	$2.29*10^{14}$	Hz
P_R	Received optical power	0.5 (R=50m)	nW
P_b	Received optical background power	10	nW
M	Avalanche gain of APD	5	
F(M)	Excess noise factor of APD	2.9	
I_d	APD dark current	20	nA
B	Electrical bandwidth	1000	Hz
T	Absolute temperature of sense resistor and LNA	300	K
R	Distance between target and LRF	1 to 50	m
F	Noise figure of LNA	2.5	dB
k	Boltzmann's constant	$1.38*10^{-23}$	JK^{-1}
m	modulation depth of laser diode intensity	70	%

Table 1

High repetition rate eyesafe rangefinders

V. J. Corcoran

Photon Interactions
5645E General Washington Drive
Alexandria, Virginia 22312

ABSTRACT

Two types of high repetition rate eyesafe rangefinders are considered. The first is an Nd:YAG laser rangefinder that operates at 1.54 microns. The second is a MIL-qualified laser diode designed for a missile.

1. INTRODUCTION

The eyesafe spectral band extends from 1.535 to 1.545 microns according the Class I Accessible Emission Limits for pulsed lasers with pulse durations of less than 100 ns. The eyesafe spectral region is not restricted to this band. It simply means that the threshold for eye damage in this spectral band is higher than for other wavelengths as indicated in Table 1. Table 1 also indicates that for pulses longer than 100 ns, the eyesafe spectral band is much broader.

Eyesafe rangefinders have been receiving increased attention in the past few years. The requirement has been generated by the need for insuring the safety of personnel using rangefinders and by the fact that lasers have become available that make eyesafe requirements more than abstract wishes. In particular, two approaches have been supported for low repetition rate rangefinders for military applications. The first uses erbium glass and the second uses a methane Raman-shifted Nd:YAG laser.

This paper is concerned with high repetition rate eyesafe rangefinders. Two different types of systems are addressed, but only one of the systems operates in the 1.535-1.545 micron spectral band, illustrating that, in fact, eyesafe operation is not limited to the 1.535-1.545 micron band. The first rangefinder considered uses a methane Raman-shifted Nd:YAG laser, which is simply a higher repetition rate of an existing system. The sedond is a laser diode rangefinder that was constructed for missiles using a pulsed laser diode transmitter. It operates at 904 nm and has been MIL qualified.

As mentioned above, one approach to eyesafe rangefinders in the 1.54 micron range is to use an Er:glass. This rangefinder is restricted to repetition rates less than five Hz by the limitations of the laser material. Above that repetition rate, however, the performance plummets.

When a 1064 nm Nd:YAG laser pumps a high pressure methane Raman shifter, the output radiation is also at 1.54 microns. This system is limited primarily by the repetition rate of the Nd:YAG laser, which can be in the Kz range. Consequently, the use of a Raman-shifted Nd:YAG laser is a virtual necessity at repetition rates of five Hz or higher for ranges typically required for a battlefield rangefinder.

For the relatively short range requirements of a missile rangefinder, eye safety can be achieved with lower peak power levels so laser diodes become usable sources as demonstrated by the missile rangefinder. Although eye safety was not a firm requirement for this device, it, nevertheless is eyesafe.

2. RAMAN-SHIFTED RANGEFINDER

A block diagram of a high repetition rate Raman-shifted rangefinder is given Figure 1. As is evident, the system is essentially like any Nd:YAG rangefinder, except for the fact that a silicon avalanche photodiode (APD) cannot be used in the receiver at 1.54 microns. The detector of choice for 1.54 micron rangefinders has been an InGaAs PIN photodiode, but the emergence of InGaAs APD's will help to improve the performance of these systems.

An Nd:YAG laser, which is Q-switched by either a dye Q-switch or a Pockels cell produces pulses that are typically 1-15 ns. Flashlamp pumped Nd:YAG lasers with repetition rates exceeding 1 KHz were demonstrated years ago (Reference 1).

When the laser pulse is generated, a fraction of the Nd:YAG radiation scattered from the optics is detected with an inexpensive silicon PIN diode. The output of the diode is used to start a clock that counts down at a 30 MHz rate. When a signal returns, the detected radiation generates a pulse that stops the count of the clock. This information is translated into a range which is indicated on the digital display.

The range resolution of the system is determined by the possible error in the last count, which depends on the clock frequency. For a 30 MHz clock the achievable resolution is given by the equation

$$
\begin{aligned}
t = 1/f &= 2*R/c \\
R &= c/2*f \\
&= 3 \times 10^{8}/2*300 \times 10^{6} \\
&= 0.5 \text{ m}
\end{aligned}
$$

In order to obtain a reasonably high conversion efficiency, a narrow linewidth laser is needed. This can be accomplished with a number of techniques including injection seeding, electronically controlled Q-switching, etc. (References 2-5). Other techniques such as a resonant Raman cell can improve the performance as well, but often the improvement goes hand-in-hand with the spectral quality of the laser (Reference 6).

Ordinarily, a Raman cell with a high average power input will have distortions or nonuniformities at the focus due to the heating of the gas that occurs at high average power inputs. In order to circumvent this problem Photon Interactions has designed two cells other than the static cell.

The first device is a circulating gas cell in which the gas is moved out of the center of the cell as illustrated in Figure 2. This cell has proved to be useful for average power inputs of two watts or less. At higher average powers it is necessary to not only flow the gas but to cool it. In this case a cell as indicated by the block diagram in Figure 3.

3. MISSILE RANGEFINDER

The MR-101 Missile Rangefinder, shown in Figure 4, is designed to fit in a missile head. State-of-the-art electronic technology is used to maintain the size and weight limitations required for a rangefinder to fit in a missile as well as to meet the technical specifications. Currently, the MR-101 weighs < 250 g, occupies a volume of < 100 cc and consumes < 0.5 W.

The MR-101 operates at a data rate of 100 Hz. It can be operated at repetition rates up to 800 Hz with increased power consumption or it can be operated at lower data rates with reduced power consumption. The current resolution is < 1 meter at all ranges, which is limited by an 8-bit A/D converter. Higher resolution, therefore, can easily be obtained in principle.

The MR-101 laser rangefinder has been environmentally tested and passed all of the tests without a failure. The tests included temperature, humidity, altitude, shock, vibration and acceleration tests according to MIL-STD-810C. In addition, the rangefinder passed RE01 and RE02 EMI tests subject to the requirements of MIL-STD-416C.

The first two MR-101 rangefinders that were tested operated to ranges > 300 m for a target such as moving automobile with a 1 cm^2 receiving aperture under sunlit conditions.

The transmitter is eyesafe not only according to the calculations that were made for zero range, but the eye safety has been verified experimentally as well.

A block diagram of the missile rangefinder is given in Figure 5. In this system an attempt is made to collimate the radiation from a single element or a multielement incoherent diode array of heterojunction GaAs laser diodes; however, the maximum aperture size is limited to one cm diameter and the focal length is also constrained because of the limited space available in the missile.

The receiving aperture is limited to one cm diameter as well. The collected radiation is spectrally filtered and relayed to a silicon APD with a built-in preamplifier. The gain of the APD is adjusted so that the background noise just exceeds the system noise.

Radiation from the laser diode array is sent through optics which take the asymmetric beam shape of the laser diode and transform it into a beam with a circular cross section. At the same time the beamwidth is narrowed.

Part of the transmitted light is detected with a silicon PIN diode. When that radiation is detected, it starts a linear voltage ramp, which is stopped when the receiver APD detects a return pulse.

By setting the charging current and capacitor value to achieve a ramp rate of, say, 4.5 volt/microsecond, the voltage at any time on the ramp can be directly converted into distance or range, e.g., for a range of 100 m, the time for the return signal is $t = 2*R/c = 2*100/3 \times 10^8 = 10^{-6}/1.5$ sec. The voltage, therefore, is $4.5/1.5 = 3$, so 3 volts = 100 meters, 6 volts = 200 meters, etc.

Since 1 volt corresponds to 100 meters, a resolution of 1 meter requires the ramp to be linear within 1/100 = 0.01 volts or 10 mv.

4. REFERENCES

1. V.J. Corcoran, R.W. McMillan, S.K. Barnoske, "Flashlamp-Pumped YAD:Nd^{+3} Laser Action at Kilohertz Rates", IEEE Jour. Quant. Electron., QE-10, 618 (1974).
2. Y.K. Park, G. Giuliana and R.L. Byer, "Single Axial Mode Operation of a Q-Switched Nd:YAG Oscillator by Injection Seeding", Jour. Quant. Electron. QE-20, 117, (1984).
3. W.D. Fountain and M. Bass, "Single-Axial-Mode Operation of a Polarization-Coupled Stable/Unstable-Resonator Nd:YAG Laser Oscillator", IEEE Jour. Quant. Electron. QE-18, 432 (1982).
4. A. Berry, D.C. Hanna and C.G. Sawyers, "High Power, Single Frequency Operation of a Q-Switched TEM_{oo} Mode Nd:YAG Laser", Opt. Comm., 40, 54 (1981).
5. D.F. Voss and L.S. Goldberg, "Simple Single Longitudinal Mode Q-Switched Nd:YAG Oscillator", IEEE Jour. Quant. Electron., QE-21, 106 (1985).
6. A.Z. Grasyuk, I.G. Zubarev and N.V. Suyazov, "Influence of Line Width of Exciting Radiation on the Gain in Stimulated Scattering", ZhETF Pis. Red. 16, 237 (1972).

TABLE 1
CLASS I ACCESSIBLE EMISSION LIMITS*
CW RADIATION

Lambda	k_1	k_2	Acc Emission Lim
810	1.59	200	1.24×10^{-4} W
904	2.49	200	1.94×10^{-4}
1060	5.00	200	3.90×10^{-4}
1540	1.0	1.0	7.90×10^{-4}

PULSED RADIATION
(t < 100 ns)

Lambda	k_1	k_2	Acc Emission Lim
810	1.59	1.0	3.18×10^{-7} J
904	1.49	1.0	4.98×10^{-7}
1060	5.00	1.0	1.00×10^{-6}
1540	100	1.0	7.90×10^{-3}

* Based on a 7 mm diameter aperture located 20 mm from the source as required by BRH.

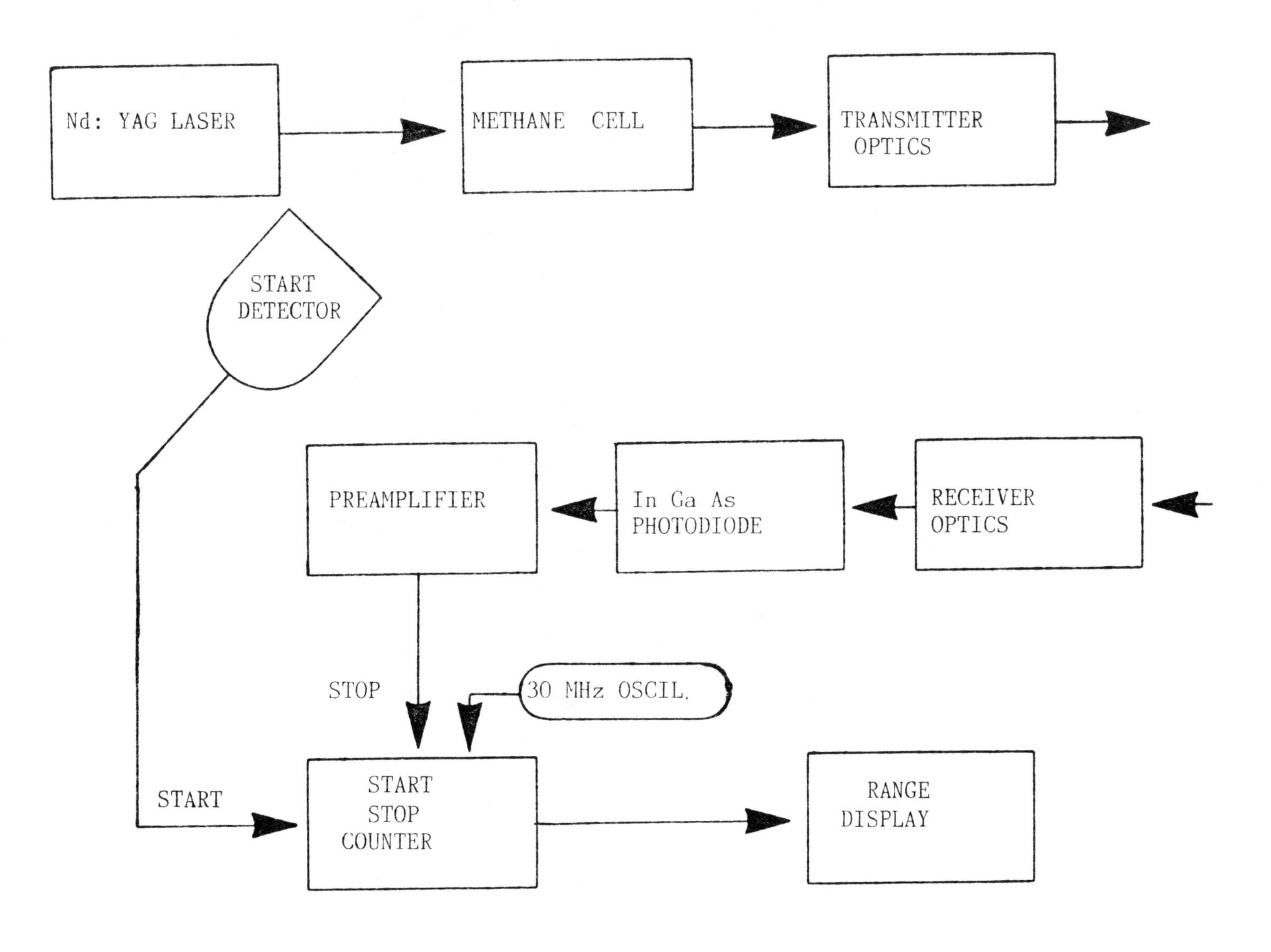

FIGURE 1. BLOCK DIAGRAM RAMAN-SHIFTED Nd:YAG RANGEFINDER

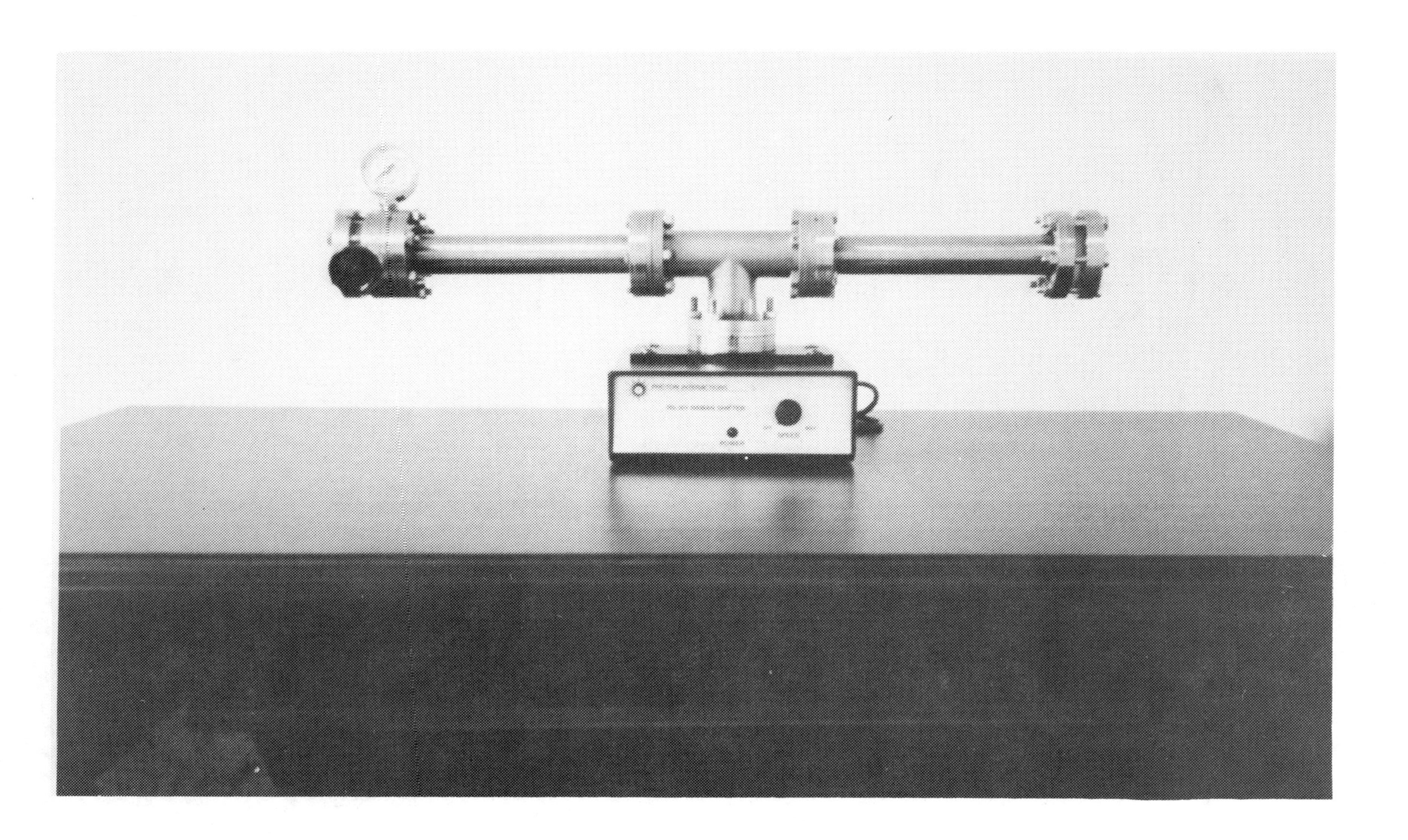

FIGURE 2. PHOTOGRAPH OF CIRCULATING RAMAN CELL

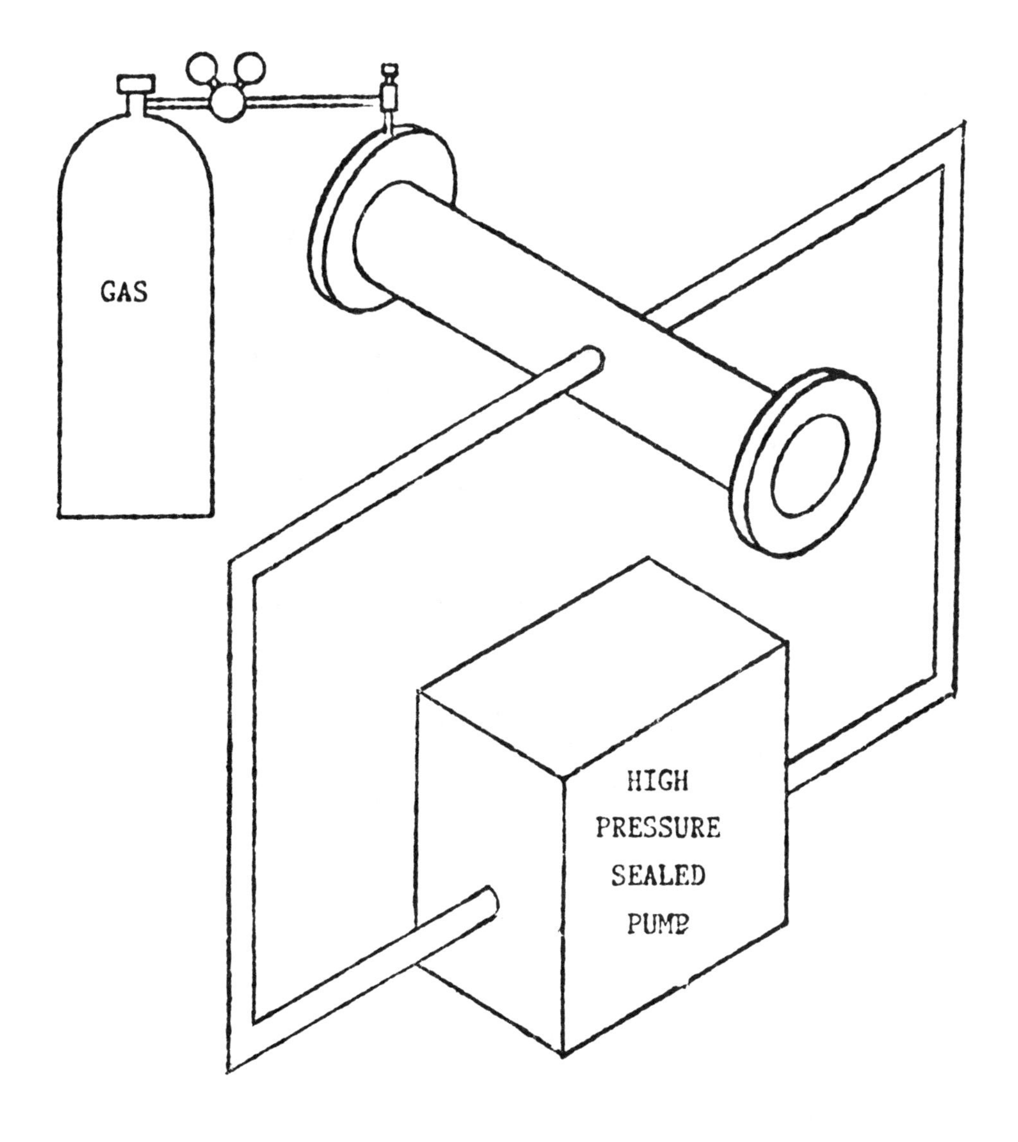

FIGURE 3. DIAGRAM OF FLOWING GAS RAMAN CELL

FIGURE 4. PHOTOGRAPH OF MISSILE RANGEFINDER

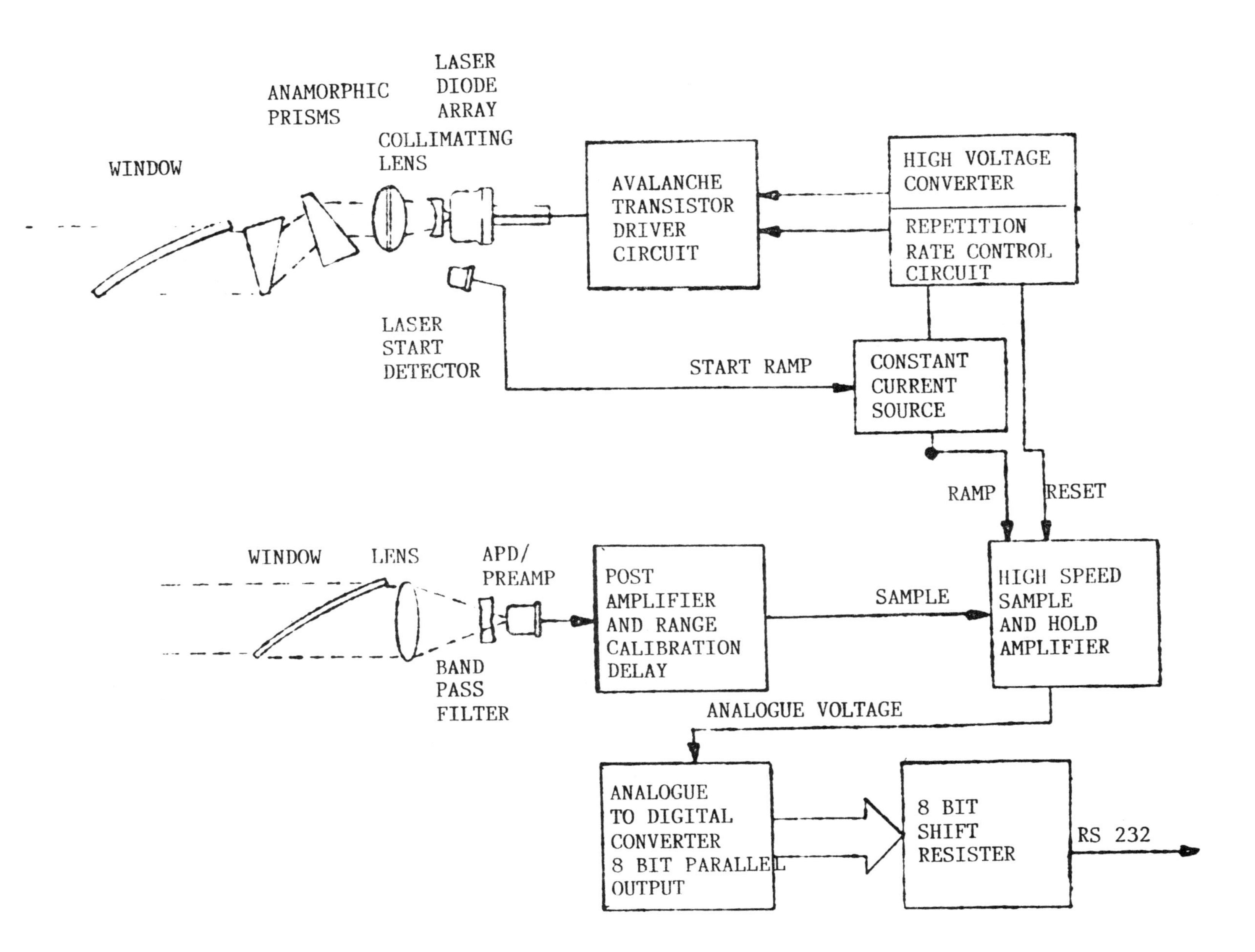

FIGURE 5. BLOCK DIAGRAM OF MIL-QUALIFIED MISSILE RANGEFINDER

AUTHOR INDEX